BIO-INSPIRED COMPUTING MODELS AND ALGORITHMS

BIO-INSPIRED COMPUTING MODELS AND ALGORITHMS

Editors

Tao Song

China University of Petroleum, China
Polytechnical Universidad de Madrid, Spain

Pan Zheng

University of Canterbury, New Zealand

Mou Ling Dennis Wong

Heriot-Watt University Malaysia, Malaysia

Xun Wang

China University of Petroleum, China

World Scientific

NEW JERSEY • LONDON • SINGAPORE • BEIJING • SHANGHAI • HONG KONG • TAIPEI • CHENNAI • TOKYO

Published by

World Scientific Publishing Co. Pte. Ltd.

5 Toh Tuck Link, Singapore 596224

USA office: 27 Warren Street, Suite 401-402, Hackensack, NJ 07601

UK office: 57 Shelton Street, Covent Garden, London WC2H 9HE

Library of Congress Cataloging-in-Publication Data

Names: Song, Tao, 1983– editor.

Title: Bio-inspired computing models and algorithms / edited by
Tao Song, China University of Petroleum, China, Pan Zheng, University of Canterbury, New Zealand, Mou Ling Dennis Wong, Heriot-Watt University Malaysia, Malaysia, and Xun Wang, China University of Petroleum, China.

Description: New Jersey : World Scientific, [2019] | Includes bibliographical references and index.

Identifiers: LCCN 2018051021 | ISBN 9789813143173 (hc : alk. paper)

Subjects: LCSH: Natural computation.

Classification: LCC QA76.9.N37 B57 2019 | DDC 006.3/8--dc23

LC record available at https://lccn.loc.gov/2018051021

British Library Cataloguing-in-Publication Data

A catalogue record for this book is available from the British Library.

For any available supplementary material, please visit
https://www.worldscientific.com/worldscibooks/10.1142/10119#t=suppl

Desk Editor: V. Vishnu Mohan

Typeset by Stallion Press
Email: enquiries@stallionpress.com

Preface

Look deep into nature, and then you will understand everything better.

Albert Einstein

This book is dedicated to a family of computer algorithms and methods that took ideas inspired by nature and biological organisms living within it. Living organisms, regardless of their sizes, never fail to amaze one with their capability to adapt and evolve in reacting to hostile environments. We have selected ten individual contributions to provide a broad view of Biological Inspired Computing. The book covers a diverse range of topics including membrane computing: a computational framework inspired by the human brain, evolutionary algorithm, self-organizing neural work, behavioural-based feature selection method, as well as DNA computing. Most of the chapters in this book also shed lights to the practicality of these algorithms through real-life application examples and case studies.

We are truly in debt to the contributors of the individual chapters who reign from a diverse geological distributions: Australia, China, India, Italy, Malaysia, South Korea, as well as Spain. We have briefly summarized the contributions of each chapters as follows.

Leporati *et al.* provide a very friendly introduction of membrane computing and properties of it as the first chapter of this book. The chapter, "A Gentle Introduction to Membrane Systems and Their Computational Properties", focuses on the basic concepts and formal definitions of membrane systems, the computational capabilities and efficiency of such systems. Interesting examples are provided to illustrate those ideas.

Their work is nicely complemented by Chapter 2 "Results on Computational Complexity in Bio-inspired Computing" by Pérez-Jiménez *et al.* This chapter examined the computational complexity in the context of membrane computing and provided a collection of interesting and insightful results.

In Chapter 3 "Integrative Approaches for Predicting microRNA Function and Prioritizing Disease-Related microRNA Using Biological Interaction Networks", Zhang and Zeng present their work on using biological interaction network to predict microRNA function and prioritize disease-related microRNA. The existing approaches and present challenges of the topic are comprehensively discussed.

The survival of fittest had always been the central theme in the theory of evolution and natural selection. Evolutionary algorithms are essentially stochastic optimisation algorithms that borrowed this idea from natural selection. In Chapter 4 "Evolutionary Algorithm for Solving Complex Multiobjective Optimization Problems", Tian and Zhang present their recent work on evolutionary algorithms in dealing complex problem with multiple optimization objectives where some objectives are counter. Several techniques for enhancing the evolutionary algorithm in solving complex MOPs are introduced.

In Chapter 5 "A Bio-inspired Clustering Model for Anomaly Detection in the Mining Industry", Chiong *et al.* introduce a bio-inspired clustering model based on the self-organizing map (SOM). The model is applied to check anomalies in the data generated by various sensors used to monitor mining machinery and aims to avoid undesired failures and costly interruptions.

A computational idea inspired by the recombinant behaviour of DNA molecules and its application is introduced in Chapter 6 "Picture Generating Models Based on the Splicing Operation". This interesting work is carried out by K. G. Subramanian *et al.* A new 2D picture array generating grammar system is proposed and its generative power is examined

Jin *et al.* put forward a thorough review of recent development in biometric field with focus on kernel learning and bag-of-minutiae applied in fingerprint recognition in Chapter 7 "Fingerprint Minutiae to Fixed-Length Bit-String: A Review of Recent Development". As a background study, the point-to-string conversion method is revisited and a summary is given on the challenges and future prospect at the end of the chapter.

Chapter 8 "Platform for Detection of Ultra-DNA Signals (10–18 mol/L) with DNA Probe and Nanomaterials" by Wang *et al.* introduces a state-of-the-art platform for detection of ultra-DNA signals (10–18 mol/L). Additionally, relevant DNA sensing technologies are discussed in this chapter.

Deep learning has been widely recognized as the recent significant breakthrough in the development of artificial neural networks. Chapter 9 "Deep Autoencoders in Pattern Recognition: A Survey" by Chen *et al.* looks into an important component, autoencoders, in deep neural network and introduces the role and significance of different categories of autoencoders with applications in pattern recognition.

The final chapter, "Genetic Algorithm: From Promiscuity to Monogamy", is another contribution in evolutionary computing, or more specifically, on genetic algorithm (GA). Lim and Tan present their investigation on the crossover operation performed in GA and convey a refreshing concept of monogamy in GA. Critically they show evidences mathematically and empirically that monogamy can be a promising way of GA implementation.

We sincerely appreciate all the contributors of this book, without whom we cannot make this volume of book successful. As editors of this text, we have been blessed by assistance from many people on all aspects of preparation of this book and fortunate to work with a group of creative and insightful people at World Scientific Publishing. These people include: Don Mak who initiated and invited us to produce this volume, Steven Patt who oversaw the entire production of this book and V. Vishnu Mohan who actively engaged with final editing and revision process of this book.

Special thanks go to *the International Membrane Computing Society* (*IMCS*) for its continuous support to the researchers working in membrane computing related research.

We also would like to thank our families for their constant support and encouragement throughout the editing and production of this book.

Tao Song
Pan Zheng
M. L. Dennis Wong
Xun Wang

May, 2019

Contents

Chapter 1

A Gentle Introduction to Membrane Systems and Their Computational Properties

Alberto Leporati*, Luca Manzoni†, Giancarlo Mauri‡, Antonio E. Porreca§ and Claudio Zandron¶

Dipartimento di Informatica, Sistemistica e Comunicazione
Università degli Studi di Milano-Bicocca
Viale Sarca 336, Building U14, 20126 Milano, Italy
**leporati@disco.unimib.it*
†luca.manzoni@disco.unimib.it
‡mauri@disco.unimib.it
§porreca@disco.unimib.it
¶zandron@disco.unimib.it

Membrane computing, known as a novel branch in computer science, has gain plenty of results in both theoretical and application levels. In this chapter, a gentle introduction to membrane systems and their computational properties is discussed. Membrane structures, the content of region, computation rules are recalled, as well as the formal definition of membrane systems is given. The computational power, as both generating and accepting deceives are shown, and the computational efficiency are discussed by solving computational hard problems by membrane systems.

1.1. Introduction

The theory of computation investigates the nature and properties of algorithmic procedures. This field emerged in the 1920s and 1930s of the 20th century from the work on the philosophy and the foundations of mathematics. Of great importance and inspiration was David Hilbert's ambitious program to "dispose of the foundational questions in mathematics once and for all" [34], that led to fundamental results in logic such as Gödel's incompleteness theorems [6], and ultimately to the birth of recursion theory (nowadays mostly referred to as computability theory) and computer science itself.

The formal notion of computability that is almost universally adopted today is due to Alan Turing, who introduced in his groundbreaking paper [32] a simple, elegant and convincing mathematical formalization of the notion of computation, as it is carried out by a human executor equipped with enough scratch paper. Turing's work showed that, as long as we accept his notion of computation, there exist well-formed mathematical questions whose answer cannot be computed. In particular, one of the main challenges of Hilbert's program, the *Entscheidungsproblem* (finding a decision procedure for the validity of statements in first-order logic) was proved to be unsolvable.

This formalization, that rapidly became known as the *Turing machine*, is still the reference model for computing devices in theoretical computer science, as it also enjoys the property of being a good model of actual electronic computers; this is also due to the fact that it was itself an inspiration for the design of automatic computing machinery [5].

With the development of computers as a technology, being able to solve a particular problem proved not to be satisfying: *fast*, *efficient* solutions are needed. This led to the development of computational complexity theory, pioneered [9] by Hartmanis and Stearns in [12], which also gives the name to the field. Identifying the notion of "efficient" with "polynomial-time computable" is due [9] to Edmonds [7], while the central question of complexity theory, whether $\mathbf{P} = \mathbf{NP}$, arose from the work of Cook [4] and Karp [16]. This question has shaped the whole development of the field and still remains open today.

However, the theory of computation is not entirely about Turing machines. Several authors sought to draw inspiration from the way nature "computes" in order to define alternative, unconventional computing models or, from the opposite point of view, to interpret natural phenomena as computation [1]. For instance, artificial neural networks [21] are inspired by the functioning of neurons in the brain, and genetic algorithms seek to solve computationally hard problems by simulating the processes of mutation, mating and natural selection. A clear example of biological inspiration is given by DNA computing, which provided an actual *in vitro* implementation of an algorithm for the Hamilton path problem [2] (the theory was initiated a few years earlier, particularly by Head [13]).

Inspired by the work on DNA computing [26], Pun introduced in 1998 (see [24]) the notion of *membrane systems*, initially called *super-cells*, and nowadays usually known as *P systems*. Here the computing device is an abstraction of biological cells. Unlike in neural networks, we do not deal with cells as atomic objects: as the name suggests, the focus is on the membranes that define the cells and their internal compartments, which work together by performing different individual functions. The chemical environment of the various compartments are described in terms of *multisets* of symbols (i.e. sets in which each symbol has a multiplicity). Another defining feature of P systems (as they were initially defined) is *maximal parallelism*: as many operations as possible are carried out simultaneously, and no part of the system remains inactive if it can carry out some part of the computation.

Although P systems have also been investigated from the perspectives of bioinformatics and systems biology, where they might be used as models of biological phenomena in computer simulations, most of the research in membrane computing has been carried out from a language-theoretic (including the seminal paper [24]), computability-theoretic and complexity-theoretic standpoint.

In the literature, many variants using various kinds of rules and different features have been considered and investigated in this respect. For example, the notion of multiset rewriting employed in the original model of the P system [24] is context-sensitive (or *cooperative*, which is the term normally used in membrane computing). Other classes of P systems, however, possess other features that make them extremely powerful from a computational perspective: as a consequence, the use of *context-free* (or *non-cooperative*) rewriting rules has also been considered. Here, for the sake of explanation, we will present a simplified model using only the basic ingredients and simple cooperative rewriting rules. After presenting some basic definitions, we will first show that such systems are Turing complete, by simulating register machines, a well-known universal computing device, and we will show that the direct simulation of Turing machines is time efficient. Then, by exploiting the possibility of creating new membranes by division of existing ones, we will show how to solve all the problems from the complexity classes **NP** and **PSPACE** in polynomial time (and exponential space) by means of membrane systems. At the end of the chapter, we will shortly discuss how cooperation can be avoided, and we will provide pointers to the literature concerning main variants considered in the framework of membrane computing.

1.2. Membrane Structures

The main feature of all variants of P systems is their *membrane structure*. Several variants have been proposed in the literature [27], but we consider here the original *cell-like* shape [24]: a hierarchy of membranes nested to an arbitrary depth. There is a clear bijective mapping between the set of membranes and the *regions* they define, delimited by the membrane itself from the outside, and any membrane immediately included in it from the inside. The outermost membrane (sometimes called the *skin*) separates the actual P system from the surrounding *environment*, which is a further implied region also playing a role in the computation. Membranes not containing further membranes inside them are called *elementary*; the others are called *non-elementary*.

The abstract shape of a membrane structure in any given instant is mathematically represented by a *rooted, unordered tree*. Here the membranes (equivalently, the regions they delimit) correspond to vertices, the outermost membrane being associated with the root of the tree, and an edge between two vertices indicates that the corresponding membranes are located one immediately inside the other; the "parent" and "child" roles are implied by the distance from the root. The leaves of the tree correspond to elementary membranes, while the internal nodes correspond to the non-elementary ones. The tree is unordered (i.e. there is no distinguishable "first" or "leftmost" child) because we do not keep track of any spatial relation between the membranes other than simple containment.[a]

Membranes are not only set apart by their position in the membrane structure, but also by the role they supposedly perform. This is represented in P systems by attaching a *label* taken from a finite set Λ to each membrane (formally, to the vertex representing it in the corresponding unordered tree). As we shall see later, different labels correspond to different sets of rules that can be applied to the membranes or their contents. Initially, each membrane must be given a different label (although two distinct labels might be associated to the same set of rules), while the process of membrane division may create multiple membranes sharing the same label.

A membrane structure is traditionally described in a "linear" notation by means of a string of balanced brackets. Given the unordered rooted tree corresponding to the membrane structure, fix an arbitrary ordering of the children of each node. The resulting ordered tree, ignoring the node labels, is then uniquely interpreted as a string of balanced brackets belonging to the language defined by the following context-free grammar, beginning with the non-terminal S_1:

$$S_1 \to [S_2] \qquad S_2 \to [S_2] \mid S_2 S_2 \mid \epsilon$$

These are all the non-empty strings of balanced brackets having a single outermost pair. The nesting of brackets is induced by the shape of the tree (and obviously corresponds to the containment relation of the original membrane structure). Finally, each bracket is subscripted by the label of the corresponding membrane. An example of membrane structure, represented in a pictorial form, as a tree, and in bracket notation, can be found in Fig. 1.1.

[a]There exist other variants of P systems, such as the *spatial* ones, where the positions of membranes and objects do actually matter [3].

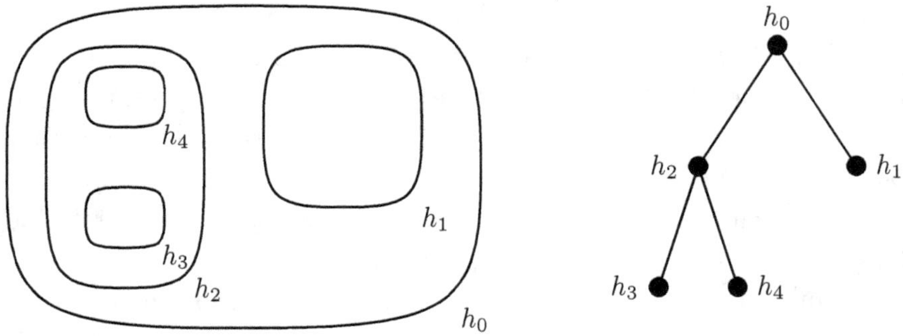

$$[[\,]_{h_1}\,[[\,]_{h_3}\,[\,]_{h_4}]_{h_2}]_{h_0}$$

Fig. 1.1. A membrane structure and two formalisms representing it.

1.3. The Content of Regions

A single molecule can be represented abstractly as a symbol taken from an alphabet Γ. However, subsets of Γ are not an adequate description of a whole chemical environment, as they do not carry any information related to the (absolute or relative) quantities of molecules of the same species. Since the concentrations (that is, amounts) of chemicals in a region do actually matter from a metabolic standpoint, we use *multisets* to describe them.

A multiset w over a set Γ is simply a function $w\colon \Gamma \to \mathbb{N}$, mapping each object to its multiplicity. Generalizations of the usual set-theoretic operations can be defined on the sets of multisets over Γ; for instance, the union of multisets w_1 and w_2 is the multiset, where for all $a \in \Gamma$, the object a has multiplicity $w_1(a) + w_2(a)$. From an algebraic point of view, the set of multisets over Γ, with respect to the operation of union, is the *free commutative monoid on* Γ. Compare this to the set $\Gamma^\star$ of strings, which is the *free monoid on* Γ: here the operation (i.e. string concatenation) is *not* commutative. When describing the chemical environment of a region, we are interested in the number of molecules, but not in any particular ordering of them.

Another equivalent characterization can be given in terms of equivalence classes of strings. For each string $u \in \Gamma^\star$ and each symbol

$a \in \Gamma$, let $|u|_a$ be the number of occurrences of a in u. Now, let u and v be strings, and let the binary relation $\simeq$ be defined by

$$u \simeq v \text{ if and only if } |u|_a = |v|_a \text{ for all } a \in \Gamma$$

It is trivial to prove that $\simeq$ is an equivalence relation, and in particular a congruence relation with respect to the operation of string concatenation. Then, the set of equivalence classes $\Gamma^\star / \simeq$ (together with the operation induced by concatenation) is isomorphic to the set of multisets over Γ (together with the operation of multiset union). This provides us with a simple way to denote multisets in writing. Let w be a multiset, and let u be any string such that $w(a) = |u|_a$ for all $a \in \Gamma$; then, we shall identify w with $[u]_{/\simeq}$ (the equivalence class of u modulo $\simeq$) and, furthermore, we shall drop the equivalence class notation, simply writing u for w, as the formal meaning of the symbol is implied; correspondingly, multiset union will be written as string concatenation. This identification justifies the following (informal) sentence, which is usually found in the P systems literature: "a multiset w will be represented as a string u, together with all its permutations", since permuting the symbols of u does not change the number of their occurrences.

1.4. Computation Rules

The elements we have described until now, i.e. the membrane structure and the multisets contained in its regions, define the instantaneous configurations of P systems. A configuration evolves by means of computation steps, where in each step some rules inspired by the biology of cells are applied.

1.4.1. *Object evolution rules*

A first kind of computation rule describes simple chemical reactions taking place inside a region. Consider, as an example, the following formula for the burning of methane as it may be described in a chemistry textbook:

$$CH_4 + 2O_2 \to CO_2 + 2H_2O$$

This notation describes how a molecule of methane (CH_4) reacts with two molecules of oxygen (O_2) yielding a molecule of carbon dioxide (CO_2) and two molecules of water (H_2O).[b] The molecules on the left-hand side of the arrow are called *reactants* or *reagents*, while those on the right-hand side are called *products*.

This kind of reaction corresponds quite closely with a limited form of *multiset rewriting*, where a certain sub-multiset is replaced by another one. The notation is also quite similar, as for multisets we shall write $u \to v$ to indicate that multiset u is replaced by v. The general notion of multiset rewriting is more abstract, and it usually disregards the law of conservation of mass. A general chemical reaction viewed as a multiset rewriting rule possesses a powerful feature from a language-theoretic perspective: it is *context-sensitive*. For instance, methane may only burn if oxygen is also present, and we require at least *two* molecules of oxygen for each molecule of methane.

The first kind of computation rules used by P systems, called *object evolution rules* or rules of type (a), are thus denoted as follows:

$$[u \to v]_h$$

This rule can be applied inside each membrane labelled by $h \in \Lambda$, having a multiset z which contains the multiset u, i.e. $u \subseteq z$. When the rule is applied, the multiset of objects defined by u is removed from z and is replaced by the multiset v.

1.4.2. *Communication rules*

The main role of cell membranes is that of delimiting and separating regions where different functions are performed. However, these

[b]The molecules involved in this reaction are, in turn, made of several atoms; however, in this discussion, we consider compound objects such as H_2O as indivisible.

distinct regions must also cooperate; for instance, the products of a chemical reaction occurred in membrane h_1 might be stored inside membrane h_2. The cell membranes are thus *selectively permeable*, allowing specific molecules to move between regions.

First, consider the case when a multiset is absorbed by a membrane. This kind of rule is called *send-in communication rule*, or type (b) rule, and is formally written as

$$u\,[\,]_h \to [v]_h$$

This rule can be applied to a membrane having label $h \in \Lambda$; the region located outside the membrane must also contain at least an instance of multiset $u \in \Gamma^*$. The effect of this rule is that the multiset u is removed from the outside region, and the multiset $v \in \Gamma^*$ appears inside the membrane. As a restriction, it is assumed that no object can be brought into the outermost membrane of the P system from the external environment.

The converse kind of rule describes the process by which a membrane expels a multiset; these are called *send-out communication rules* or type (c) rules:

$$[u]_h \to [\,]_h\,v$$

This rule can be applied to a membrane labelled by $h \in \Lambda$ containing an instance of the multiset u. By applying this rule, the system removes u from the membrane, and places the multiset $v \in \Gamma^*$ into the region immediately outside the membrane.

1.4.3. *Division rules*

We have already seen that the membrane structure of a cell need not remain unchanged during its entire lifespan. In fact, new membranes can be created. The quintessential example is given by the process of mitosis, whereby a whole copy of the original cell is produced by division. The process of division also occurs *internally* in cells (that is, without involving the outermost cell membrane), as mitochondria also divide by binary fission [17]. In general, division allows the creation of new "processing units" when the need arises.

The corresponding *division rules*, or type (d) rules, have the following form:

$$[u]_h \to [v]_h \; [w]_h$$

A rule of this kind can be applied to a membrane labelled by $h \in \Lambda$ and containing at least one instance of the multiset $u \in \Gamma^*$. Its effect is as follows: a whole new copy of the membrane is created and placed into the same region as the original one; the contents are also copied, with the exception of the instance of u mentioned above, which is replaced by an instance of $v \in \Gamma^*$ in one of the two resulting membranes and by $w \in \Gamma^*$ in the other one. This kind of rule cannot be applied to the outermost membrane.

Notice that the membrane h can be elementary or non-elementary: the difference is usually underlined by calling the corresponding rules *elementary membrane division* and *non-elementary membrane division*, respectively. In case h is a non-elementary membrane, all the membranes contained in it (as well as their content) are copied in both of the new instances of h.

1.5. A Formal Definition

We are now able to give a precise definition of our model of computation of interest.

Definition 1.1. A *P system* of initial degree $d \geq 1$ is a structure $\Pi = (\Gamma, \Lambda, \mu, w_{h_1}, \ldots, w_{h_d}, R)$ where

- Γ is the alphabet,
- Λ is the finite set of labels,
- μ is the membrane structure, a rooted undirected tree of d nodes labelled in a one-to-one way by elements of Λ,
- $w_{h_1}, \ldots, w_{h_d}$, with $h_1, \ldots, h_d \in \Lambda$, are multisets over Γ,
- R is a finite set of rules.

For each $h \in \Lambda$, the multiset w_h describes the objects initially located in the region delimited by the membrane labelled by h.

1.5.1. *Configurations and computations*

An instantaneous *configuration* of Π is given by the rooted unordered tree describing its current membrane structure, augmented as follows: each node is labelled by a couple (h, w) where

- $h \in \Lambda$ is the label of the corresponding membrane,
- w is the multiset over Γ contained in the membrane.

A computation step changes the current configuration by applying the rules in a *maximally parallel way*, according to the following principles.

- Each membrane can be subject to at most one communication or division rule per step. Any number of evolution rules can be simultaneously applied inside the same membrane, except for the restrictions on objects described below.
- Each object can be subject to at most one rule per step.
- A maximal combination of rules must be applied during each step. Each object that appears on the left-hand side of an applicable evolution, communication or division rule must be subject to exactly one of them. We say that a rule is "applicable" if the label of the membrane containing the object corresponds to those appearing on the left-hand side of the rule. The only objects that do not evolve are those associated to no applicable rule (or to no rule at all).
- Each membrane that appears on the left-hand side of an applicable (i.e. having the correct labels on the left-hand side) communication or division rule must be subject to exactly one of them. The only membranes that do not evolve are those associated to no applicable rule (or to no rule at all).
- When several maximally parallel combinations of rules can be applied at the same time, a non-deterministic choice between them

is made. As a consequence, in the general case, multiple possible configurations can be reached by performing a computation step.

- After a legitimate maximally parallel combination of rules has been chosen (i.e. every membrane and object that can evolve has been assigned to a rule), the rules are applied in a logical bottom-up (depth-first) way, from the elementary membranes towards the outermost one. Inside each membrane, in turn, first all object evolution rules are applied, then the remaining communication or division rule.

The *maximal parallel* way of applying the rules may be easily visualized as follows. Note that this is not the only way used in the literature to apply the rules: other common strategies are *sequential* application and *minimal parallelism*, whose details of operation can be found in [27]. Assume that a given region of a P system contains a multiset u of objects and a set R of rules. Each of these rules is either *enabled* by the objects in u (that is, the rule could in principle be applied to such objects) or not enabled, depending upon whether u contains the multiset of objects that appears in the left-hand side of the rule. Choose one of the enabled rules, and mark as used the objects of u that appear in the left-hand side of the rule. Now, repeat the process with the remaining (unmarked) objects of u: select one of the rules enabled by them, and mark the objects used by the rule. Repeat the process until all objects are marked as used, or the remaining objects do not enable any rule.

Since at each step one rule is chosen in a non-deterministic way, the same initial multiset u of objects may lead to several different choices of the rules to be applied. For this reason, in the literature, it is usually told that the rules are applied in a *non-deterministic* and *maximally parallel* way. Notice that no rule is actually applied until all the rules have been selected; they are then applied in parallel, each one consuming the objects mentioned in its left-hand side, and producing all the objects listed in the right-hand side. The selection process together with the parallel application of all selected rules, in each of the regions determined by the membrane structure, constitutes a single computation step of the P system. A global clock is assumed, that synchronizes the operation of all the regions of the system; each computation step takes a single clock tick.

Let us also note that maximal parallelism does not imply that the maximal number of rules which can be applied with the objects in u will be selected. Instead, maximal parallelism means that no additional rule should be applicable in the same step to idle objects of u. The difference between the two approaches can be seen through a simple example.

Example 1.1. Assume that a region contains the following set of rules:

$$r_1 : ab \to v_1 \qquad r_2 : c \to v_2 \qquad r_3 : bc \to v_3$$

$$r_4 : a^3c^2 \to v_4 \qquad r_5 : ad \to v_5$$

and the multiset of objects $u = a^3b^2c^2$ (which is a shorthand for *aaabbcc*, or any other permutation of this string).

Then, the following multisets of rules can be obtained through the selection process described above and are indeed maximally parallel:

$$\{(r_1, 2), (r_2, 2)\}$$

$$\{(r_3, 2)\}$$

$$\{(r_1, 1), (r_2, 1), (r_3, 1)\}$$

Here, the notation (r_i, n) indicates that the multiplicity of rule r_i is n, that is, rule r_i has been selected n times. It is easily seen that the fact of being maximally parallel is not related in any way with the number of rules which is selected to be applied.

A *halting computation* of Π is a sequence of configurations $(\mathcal{C}_0, \ldots, \mathcal{C}_k)$ starting from the initial one such that every $\mathcal{C}_{i+1}$ is reachable from $\mathcal{C}_i$ by performing a single computation step, and no rule at all can be applied in $\mathcal{C}_k$. If Π never reaches a halting configuration, the result is an infinite, non-halting computation $(\mathcal{C}_i : i \in \mathbb{N})$.

In the following sections, we show that the model of P systems we have defined is able to

- simulate register (counter) machines and Turing machines, hence P systems constitute a universal (in the sense of Turing-complete) model of computation;
- solve **NP**-complete problems (and thus all problems in the class **NP**) in polynomial time (and exponential space) by exploiting *elementary* membrane division rules to simulate the computations of non-deterministic Turing machines;
- solve all problems in the complexity class **PSPACE**, in polynomial time and exponential space by using division rules for *non-elementary* membranes to simulate the computations of *alternating* Turing machines.

We just recall that **NP** is the class of decision problems (equivalently, languages) which can be solved (respectively, accepted) by non-deterministic Turing machines working in polynomial time with respect to the input size. Similarly, **PSPACE** is the class of decision problems (respectively, languages) that can be solved (respectively, accepted) by deterministic or non-deterministic Turing machines operating in polynomial space with respect to the input size. An excellent reference for the computational properties of these (as well as many others) complexity classes is [23].

1.6. Universality of Multiset Rewriting

Cooperative (or context-sensitive) multiset rewriting rules $u \to v$, even using only two symbols on the left-hand side, are powerful enough to reach computational universality. This result is actually simple enough to work as an introductory example.

Let us consider register machines, also called counter machines or program machines, as described by Minsky [22]. This model, inspired by the working of electronic computers, is able to compute any recursive function on natural numbers. A register machine consists of a finite number of registers $r_1, \ldots, r_n$, each one containing a non-negative integer (i.e. a natural number). The input of the register machine is located in register r_1, while the remaining registers are initially null (i.e. they contain zero). The behaviour of the counter machine is described by a finite sequence of instructions, uniquely labelled by natural numbers, and a program counter stores the label of the instruction currently being executed. The program counter is, by convention, initially null.

The variant of the register machine we consider here only has two kinds of instructions. The first kind is an increment instruction, which is denoted by the syntax

$$i\colon \text{inc}(r), j$$

Here, i is the label of the instruction, r is the name of a register and j is an instruction label. The execution of this instruction causes the increment by one of the value of the register r, and changes the value of the program counter to j.

The second type of instruction is a conditional decrement with the syntax

$$i\colon \text{dec}(r), j, k$$

where i is the label of the instruction, r is a register, and j and k two further instruction labels. The execution of this instruction depends on the current value of register r. If this is greater than zero, then it is decremented by one and the computation proceeds with the instruction labelled by j; otherwise, the register remains null and the program counter is set to k.

The computation of the register machine begins with the initial value of the program counter and proceeds by updating it according to the current instruction. The computation halts when the program counter reaches a value not associated with any instruction; alternatively, the computation can go on forever if this never happens. Halting computations can be considered to produce, as output, the final value of register r_1.

An instantaneous configuration $(i, x_1, \ldots, x_n)$ of the register machine consists of the value of the program counter i and the values $x_1, \ldots, x_n$ of the registers $r_1, \ldots, r_n$. Such configuration can be encoded as a multiset over the alphabet

$$\{p_i : i \text{ is a label appearing in the program}\} \cup \{r_1, \ldots, r_n\}$$

where the multiplicity of r_j is exactly the value x_j of register r_j for $1 \leq j \leq n$, and exactly one of the p_i's occurs and with multiplicity one, representing the current value of the program counter. Hence, an arbitrary configuration $(i, x_1, \ldots, x_n)$ of the register machine is encoded as $p_i\, r_1^{x_1} \cdots r_n^{x_n}$.

The instructions of the register machine are simulated by multiset rewriting rules operating in the maximally parallel way. Each increment instruction $i\colon\ \mathrm{inc}(r), j$ is implemented as a single rewriting rule

$$p_i \to p_j r$$

which simultaneously replaces the program counter symbol with the updated one, and increments by one the multiplicity of symbol r as required.

A conditional decrement instruction $i\colon\ \mathrm{dec}(r), j, k$ requires checking whether symbol r occurs with multiplicity zero or greater than zero; this can be implemented by using appropriately synchronized cooperative rewriting rules [10]. First of all, symbol p_i is rewritten by

$$p_i \to p_i'\, d_r$$

where symbol d_r will try to delete an occurrence of r, while p_i' waits one step in order to allow the following decrement to occur:

$$d_r r \to d_r' \qquad p_i' \to p_i''$$

Notice that the first rule can only be applied when at least one instance of the symbol r occurs: in that case, symbol d_r' is produced; otherwise, symbol d_r remains inert and is carried on to the next configuration. Symbol p_i'' is able to discriminate between these two cases by applying one of the following mutually exclusive rules:

$$p_i'' \, d_r' \rightarrow p_j \qquad p_i'' \, d_r \rightarrow p_k$$

The rule that is actually applied produces the correct symbol encoding the updated value of the program counter.

The rules described above, applied to the encoding of a configuration of the register machine, produce (in up to three steps) the encoding of the next configuration, obtained by executing the prescribed increment or decrement operation. This procedure halts if and when the simulated register machine halts; in that case, the output of the register machine can be recovered from the multiplicity of symbol r_1.

1.7. Efficient Simulation of Polynomial-Time Turing Machines

Although cooperative multiset rewriting rules suffice to attain universality, the register machines simulated in the previous section are, in general, exponentially slower than the reference model for efficient computation, namely Turing machines [8].

Turing machines themselves can be simulated by multiset rewriting systems when the space employed (i.e. the tape length m) is *a priori* known. The symbols in a multiset do not, by themselves, possess the linear ordering of the symbols on the tape of a Turing machine. This can be simulated by indexing the symbols of the tape alphabet Σ of the Turing machine with integer subscripts ranging from 1 to m, the space bound previously assumed. This allows us to encode the first m tape cells (including the blank ones) as a multiset of m symbols. The current state q of the Turing machine, ranging over the set of states Q, and the current position i of the tape head (also ranging from 1 to m) can be encoded by an extra symbol q_i, which contains both pieces of information. For instance, a sample Turing machine configuration is encoded in Fig. 1.2.

Let $\delta\colon Q \times \Sigma \rightarrow Q \times \Sigma \times \{-1, 0, +1\}$ be the transition function of the Turing machine. Each transition $\delta(q, a) = (r, b, d)$ is implemented

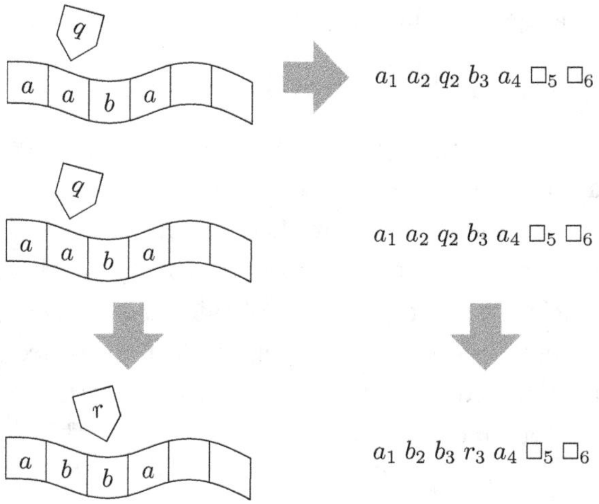

Fig. 1.2. On the left, a Turing machine performing the transition $\delta(q, a) = (r, b, +1)$ and, on the right, the corresponding rewriting step between two multisets encoding the configurations of the Turing machine.

by a rewriting rule for each position $1 \leq i \leq m$ of the tape:

$$q_i\ a_i \rightarrow r_{i+d}\ b_i$$

These rules simultaneously update the state and head position of the simulated Turing machine, and the content of the cell previously located under the tape head, as exemplified in Fig. 1.2.

This simulation is carried out efficiently, actually in "real time" (each step of the Turing machine is simulated by a single rewriting step), and shows that multiset rewriting can be efficient, whenever the space bound is known. This means that, unlike the register machine simulation (where a single rewriting system simulates the given register machine on all possible inputs), a *family* of rewriting systems is needed, depending on the space bound (which is, ultimately, a function of the length of the input string).

The rewriting systems of the family differ only with respect to the subscripts of its symbols, since the rules simulating the transitions of the Turing machine are simply replicated for each cell index. This means that the family is *uniform*: there exists a single algorithm (and an *efficient* one on top of that) for constructing the mth member of the family, where m is the length of the tape to be simulated. This echoes similar constructions for models of computation, such as families of Boolean circuits [33], where the shape of each circuit does, of course, depend on the number of input bits.

1.8. Simulating Non-determinism with Membrane Division

Using membranes allows us to partition the space in multiple regions evolving in parallel, possibly according to distinct sets of rules. For instance, the rewriting rule $u_1 \to v_1$ might only be associated with membrane label h, while the rewriting rule $u_2 \to v_2$ is associated with label k. We denote such region-specific rules with the syntax

$$[u_1 \to v_1]_h \qquad [u_2 \to v_2]_k$$

A mechanism to allow distinct regions to exchange objects is given by *communication rules*, either sending multisets towards the region outside a membrane h or towards an internal membrane k:

$$[u_1]_h \to [\,]_h\, v_1 \qquad u_2\, [\,]_k \to [v_2]_k$$

These rules send out the objects in the multiset u_1 (respectively, send in u_2) towards the target region, while simultaneously rewriting the multiset into v_1 (respectively, v_2). Notice that a send-in communication rule is a potential source of non-determinism, in the case that the region where the rule is applied contains two or more membranes with label k. By using communication rules, the multisets can be moved around the P system, allowing distinct processing to occur, such as conditional behaviour or "subroutines".

The role of membranes becomes dramatic when new membranes can be created during the computation. The most common approach is *membrane division*, which mimics the biological process of binary fission (occurring, for instance, in cell mitosis). A simple kind of membrane division rule has the form

$$[u]_h \to [v]_h \; [w]_h$$

where h is a membrane label and u, v and w are multisets. This rule is applicable to a region delimited by a membrane with label h that contains u as a submultiset. Membrane h is duplicated, its content is replicated inside both resulting membranes, with the exception that multiset u is rewritten into v in one membrane and into w in the other one. This allows us to differentiate the two membranes, which can then operate in parallel on distinct data.

If further rules are simultaneously applicable inside the membrane h, the usual semantics [25] requires the division process to logically occur after the remaining rules have been applied, although this only requires a single time step. This way, the only difference between the two resulting copies of membrane h is given by the two multisets v and w on the right-hand side of the division rule. If membrane h contains further membranes, the whole subtrees are replicated when the membrane divides.

The power of membrane division stems from the possibility to repeatedly apply division rules, creating in polynomial time an exponential number of membranes working in parallel. This allows us, for instance, to simulate non-determinism using parallelism.

Let us consider the Turing machine simulation of Section 1.7, and extend it to non-deterministic Turing machines, allowing the transition function δ to assume (without loss of generality) one or two values for each input pair (q, a). Let us place the multiset encoding the instantaneous configuration of the Turing machine inside a membrane h. A deterministic transition $\delta(q, a) = \{(r, b, d)\}$ is simulated as before by the rules

$$[q_i \; a_i \to r_{i+d} \; b_i]_h \quad \text{for } 1 \leq i \leq m$$

On the other hand, a non-deterministic transition $\delta(q, a) = \{(r, b, d), (s, c, e)\}$ can be simulated by dividing the membrane, and carrying on the two possible computations in parallel inside the two resulting membranes:

$$[q_i\ a_i]_h \to [r_{i+d}\ b_i]_h\ [s_{i+e}\ c_i]_h \quad \text{for } 1 \le i \le m$$

When the simulated Turing machine reaches a halting state, either an accepting state q_{yes} or a rejecting state q_{no}, the corresponding object $q_{yes,i}$ or $q_{no,i}$ (for some $1 \le i \le m$, corresponding to the position where the tape head is located upon halting) is sent out by a communication rule

$$[q_{yes,i}]_h \to [\,]_h\ q_{yes} \quad \text{for } 1 \le i \le m$$

$$[q_{no,i}]_h \to [\,]_h\ q_{no} \quad \text{for } 1 \le i \le m$$

The final result of the simulated Turing machine is given by the multiplicity of the object q_{yes} in the outside region when the P system halts: acceptance if there is at least one instance of q_{yes} (corresponding to at least one accepting computation of the Turing machine) and rejection otherwise (if all simulated computations were rejecting).

Figure 1.3 shows the correspondence between the non-deterministic choices of a Turing machine and the membranes obtained by division and operating in parallel in the P system simulating it.

1.9. Simulating Alternation Using Non-elementary Membrane Division

The simulation of non-deterministic Turing machines in Section 1.8 only requires division rules for elementary membranes. The efficiency of P systems increases[c] by exploiting the division of non-elementary membranes, which replicates whole subtrees of the membrane structure. This makes it possible to simulate not only non-determinism, but also *alternation*, which allows Turing machines to characterize the complexity class **PSPACE** in polynomial time [23].

[c]Under standard complexity theory assumptions, namely **NP** $\neq$ **PSPACE**.

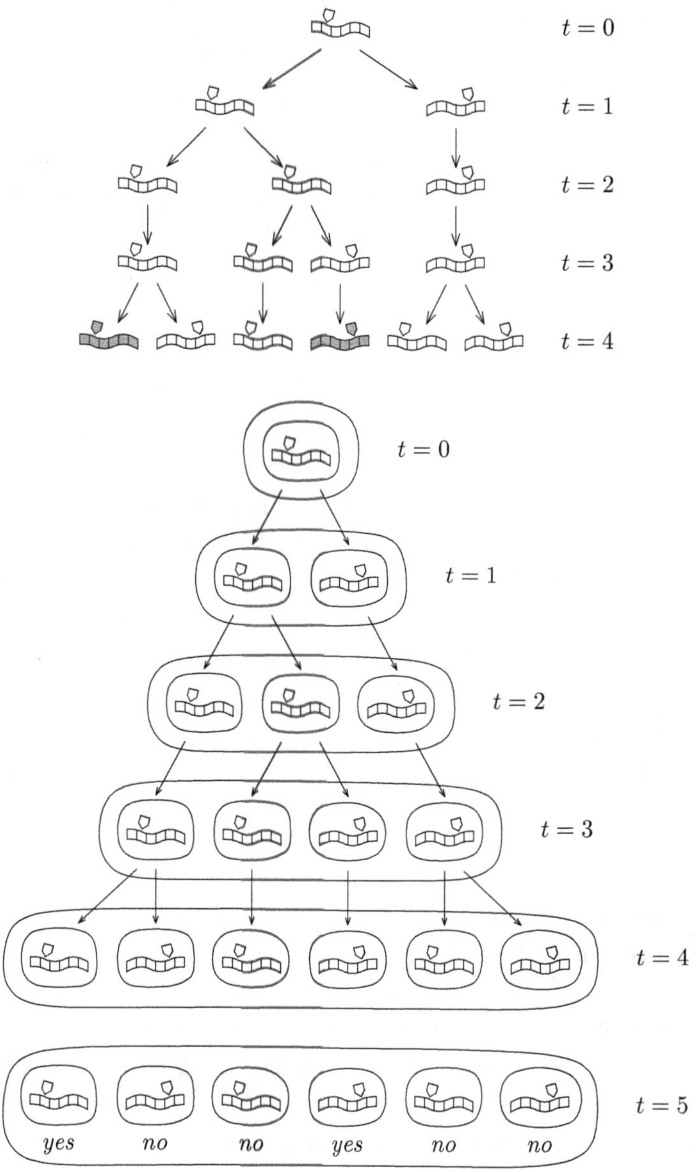

Fig. 1.3. The computation tree of a non-deterministic Turing machine, with the accepting computations highlighted in grey (top), and the corresponding parallel computation of a P system simulating it (bottom). Each non-deterministic transition of the Turing machine is simulated by a membrane division, as highlighted by the corresponding arrows of the two diagrams. In the last computation step of the P systems ($t = 5$), the results of all simulated computations are sent out to the outermost membrane.

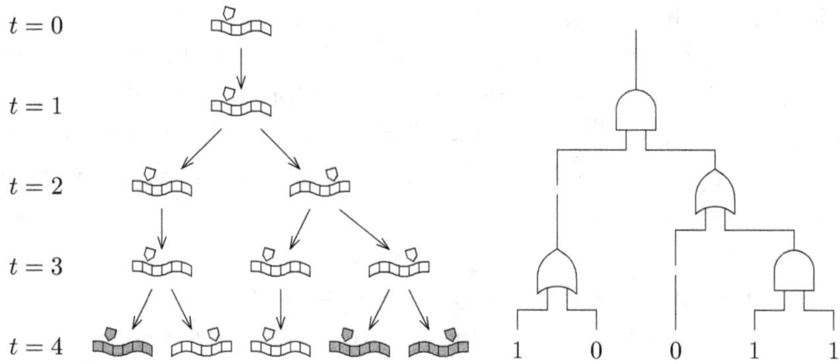

Fig. 1.4. The computation tree of an alternating Turing machine (left), with the kind of state, existential or universal, represented by an isomorphic circuit with OR and AND gates, respectively (right).

Recall that an alternating Turing machine has a set of states Q partitioned into *existential* and *universal* states. A configuration leads to acceptance if either

- the state of the machine is accepting, or
- the state of the machine is existential, and at least one successor configuration leads to acceptance, or
- the state of the machine is universal, and all successor configurations lead to acceptance.

The Turing machine is said to accept its input if the corresponding initial configuration leads to acceptance and to reject otherwise. A sample alternating computation tree is shown in Fig. 1.4.

Notice that a standard non-deterministic Turing machine is simply an alternating Turing machine where all states are existential, and thus there is *no* alternation between existential and universal.

Let us consider an alternating Turing machine M working on input x; suppose, without loss of generality [23], that all computations have the same length t. We simulate the machine by means of a P system by generating (via membrane division rules) a membrane structure isomorphic to the computation tree of M. The initial membrane structure consists of $t+1$ linearly nested membranes; since the rules will be the same for each membrane, these can be assumed to be all identically labelled by h. Notice that this last assumption is not strictly compatible with the formal definition of membrane structure, but can be simulated by replicating all rules for each membrane label; we can thus safely make this assumption without changing the computing power of the model. The encoding of the configuration of M (as described in Section 1.8) is initially located inside the outermost membrane.

The simulation of each computation step of M is similar to the one illustrated in Section 1.8, but before performing this simulation, the encoding of the configuration of M is moved one level deeper in the membrane structure, while leaving behind an object (q, a) representing the current state of the Turing machine and the object under its tape head. This process begins with the following rule:

$$[q_i \, a_i \to (q, a) \, q_{i,1} \, a_i]_h \quad \text{for } q \in Q, a \in \Sigma \text{ and } 1 \le i \le m$$

The object $q_{i,j}$ with $1 \le j \le m$ brings the object x_j encoding the symbol on the jth tape cell of M into the internal membrane:

$$q_{i,j} \, x_j \, [\,]_h \to [q'_{i,j} \, x_j]_h \quad \text{for } q \in Q, 1 \le i \le m \text{ and } 1 \le j \le m$$

If there remain objects to be brought in (i.e. if $j < m$), the object $q'_{i,j}$ is sent out with an incremented subscript:

$$[q'_{i,j}]_h \to [\,]_h \, q_{i,j+1} \quad \text{for } q \in Q, 1 \le i \le m \text{ and } 1 \le j < m$$

Otherwise, the configuration of M has been completely moved one level deeper in the membrane structure, while leaving the object (q, a) outside. The current computation step of M can now be simulated. If this is a deterministic transition $\delta(q, a) = \{(r, b, d)\}$, then the P system applies one of the rewriting rules

$$[q'_{i,m} \, a_i \to r_{i+d} \, b_i]_h \quad \text{for } 1 \le i \le m$$

If the transition is non-deterministic, such as $\delta(q, a) = \{(r, b, d), (s, c, e)\}$, then the P system applies a membrane division rule, which is non-elementary in all computation steps but the last one:

$$[q'_{i,m}\ a_i]_h \rightarrow [r_{i+d}\ b_i]_h\ [s_{i+e}\ c_i]_h \quad \text{for } 1 \leq i \leq m$$

Any subsequent computation steps are simulated recursively in the same way, at deeper and deeper levels of the membrane structure.

When a computation of the Turing machine reaches an accepting state q_{yes} or a rejecting one q_{no} in the last computation step, the corresponding object is sent out as *yes* or *no*, respectively:

$$[q_{yes,i}]_h \rightarrow [\,]_h\ yes, \quad [q_{no,i}]_h \rightarrow [\,]_h\ no \quad \text{for } 1 \leq i \leq m$$

The membrane immediately outside receives either one or two objects yes, no from its child (respectively, children) membranes, depending on whether the corresponding transition was deterministic or non-deterministic. The object (q, a) that was left behind when that transition was simulated stores two pieces of information: whether the transition was deterministic, and whether the state q of M at the time was existential or universal. This process is depicted in Fig. 1.5.

If the transition was non-deterministic and the state existential, then the membrane receives two objects. The configuration previously located inside that membrane leads to acceptance if at least one of the two result objects is *yes*. This information is recursively propagated back towards the outermost membrane by using the rules

$$[(q,a)\ yes\ yes]_h \rightarrow [\,]_h\ yes \quad [(q,a)\ yes\ no]_h \rightarrow [\,]_h\ yes$$
$$[(q,a)\ no\ no]_h \rightarrow [\,]_h\ no$$

that is, the two objects are combined by logical disjunction, corresponding to an existential non-deterministic choice.

Dually, if the transition was non-deterministic and the state universal, the two result objects are combined by logical conjunction,

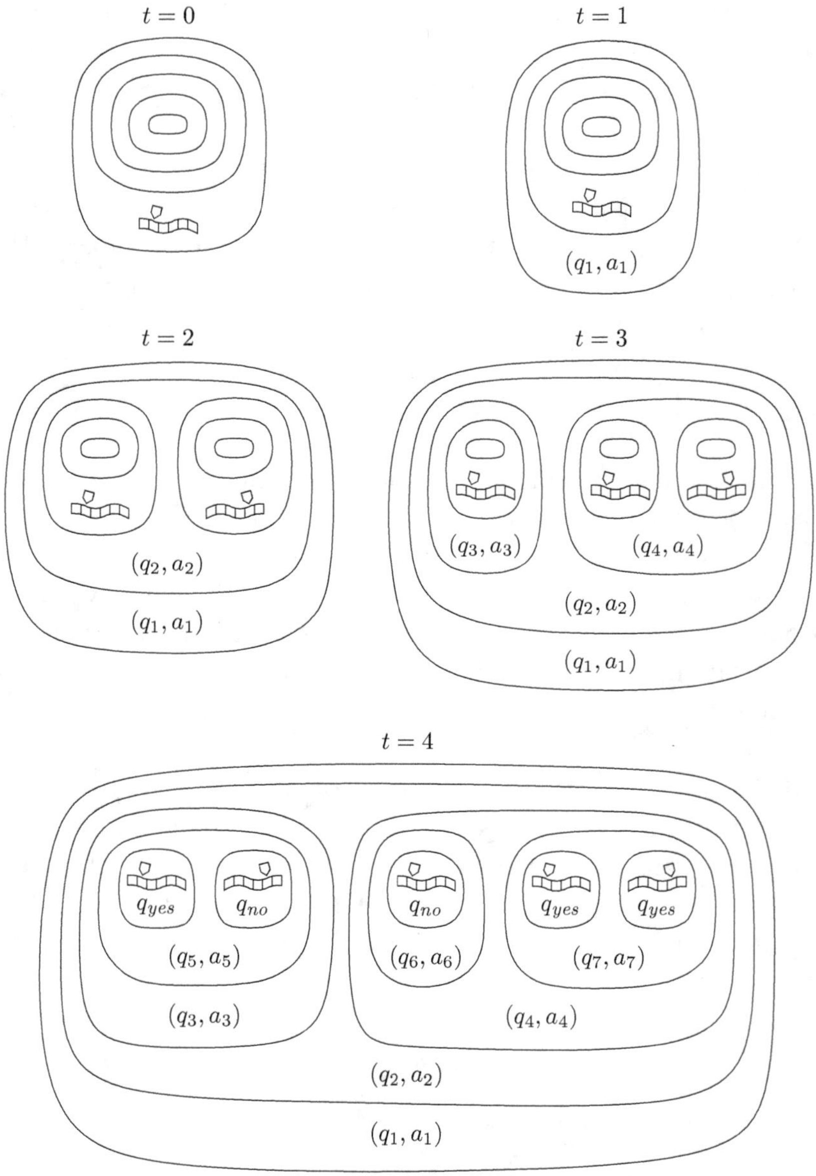

Fig. 1.5. The computation tree of Fig. 1.4 is simulated by the isomorphic membrane structure of a P system with membrane division (bottom, $t = 4$), which is constructed by using send-in and membrane division from an initially linear membrane structure ($t = 0$). The internal membranes contain encodings of configurations of the Turing machine, while each of the external ones contain an object storing the information needed to combine (by OR or AND) the results of the subcomputations being simulated by its children membranes.

corresponding to a universal non-deterministic choice:

$$[(q,a)\ yes\ yes]_h \to [\,]_h\ yes \quad [(q,a)\ yes\ no]_h \to [\,]_h\ no$$

$$[(q,a)\ no\ no]_h \to [\,]_h\ no$$

If the transition was instead deterministic, then the existential or universal nature of the state is immaterial, and the unique result object is simply propagated outside:

$$[(q,a)\ yes]_h \to [\,]_h\ yes \quad [(q,a)\ no]_h \to [\,]_h\ no$$

A simple inductive reasoning shows that the *yes* or *no* object sent out from the outermost membrane corresponds to the actual result of the simulated alternating Turing machine.

1.10. Conclusions and Pointers to the Literature

In this chapter, we tried to give a gentle introduction to P systems, using a simplified cell-based model. Indeed, since their first appearance in the literature [24], it has been clear that P systems constitute more a *framework* to define classes of abstract models of computation, rather than a specific model. A number of variants of P systems, and of models of computation inspired by them, has been proposed in the last few years.

A first class of these variants is obtained by removing or adding computational features to the cell-based model we have described above. So, for example, when investigating computability and computational complexity issues, a common observation made in the membrane computing community is that cooperative rules, having two or more objects in their left-hand side, are too powerful. Indeed, as we have shown above, it is very easy to obtain the computational power of register machines and of Turing machines (hence, universality) by means of a simple set of rules, and it is also easy to solve computationally difficult problems in polynomial time. Among the ways proposed in the literature to avoid cooperative rules, one of the most successful (in terms of number of papers published) is endowing membranes with an electric charge, which can be either positive (denoted with +), negative (−) or null (0). The resulting model is called P systems with *active membranes*. The rules which can be used in this model are the same as those defined in Section 1.4 with two important differences:

- All the membranes mentioned in each rule have an electrical charge. This allows us to partition the set of rules associated with the region enclosed by a membrane into classes, each one composed of rules which can only be applied under a given charge.
- The left-hand side of every evolution rule contains just one object. No constraints are instead imposed on the multiset of objects that these rules can produce on their right-hand side.

Adding electrical charges to membranes allows one to *move* cooperation from multisets of objects to single objects and membranes. With this limited version of cooperation, it clearly becomes more difficult to both obtain the computing power of Turing machines and solving **NP**-complete [25] or **PSPACE**-complete [30] problems. However, by using standardized constructions [19], it actually becomes much easier to design such systems. An extended literature exists on membrane systems with active membranes, especially related with the capability of solving computationally hard problems: the reader may consult the references given at the end of this chapter to find good pointers to start with.

Another example of cell-based membrane systems is *P systems with symport/antiport rules* [28], in which the rules are cooperative and are inspired from the coordinated transport of substances across the membranes. Indicated with x and y two multisets of objects, the rules may be of the following forms:

- (x, in): the objects of x are moved into the current region (where the rule is defined) from the surrounding region;
- (x, out): the multiset of objects x is moved to the surrounding region;

- $(x, in; y, out)$: the objects of x are moved from the surrounding region into the current region, and simultaneously the objects of multiset y are moved in the opposite direction.

The first two kinds of rules are called *symport* rules, whereas the rule $(x, in; y, out)$ is called an *antiport* rule. Even inside the class of cell-based symport/antiport P systems, some variants have been defined and their computational properties have been investigated; we refer the interested reader to [11] and to [27, Chapter 5].

Cell-based P system is not the only model which has been proposed in the literature. By neglecting the internal structure of cells, and focusing instead the attention on the connections between different cells, *tissue P systems* can be defined [20]. In this model, the cells are represented as the nodes of a graph, whose arcs indicate the connections between the cells, and thus provide a way to represent intercellular communication. Such a communication occurs via rules that closely resemble the above-mentioned antiport rules, and consists of an exchange of substances between adjacent cells. The interested reader can find further information in [27, Chapter 9]. Also in this case, several variants of tissue P systems have been defined: in the original model [20], the cells had an internal state, a feature which had been dropped very soon since in the computability and computational complexity setting, it is considered too powerful. On the other way, cell division rules have been added in order to allow the systems to solve computationally difficult problems [29], just like it happens with P systems with active membranes. However, the lack of internal structure prevents them from being able to solve **PSPACE**-complete problems; perhaps surprisingly, they turn out to be able to solve in polynomial time exactly the problems in $\mathbf{P}^{\#\mathbf{P}}$, the class of problems solved by Turing machines with oracles for counting problems [18].

Tissue P systems have also inspired *spiking neural* (SN) P systems [15], in which the graph of cells represents a neural network, on its turn, inspired by the structure and functioning of living nervous systems. One peculiarity of SN P systems is that they use only one kind of object, called the *spike*. Each cell — called *neuron* in this model — contains a set of rules, which may be of two different types: *spiking* rules, that send a predefined number of spikes to all neighbouring neurons, and *forgetting* rules, that delete a predefined number of spikes from the neuron itself. Spiking rules are enabled according to the number of spikes currently contained in the neuron; in fact, each rule has an associated regular expression which determines a regular set of non-negative integer numbers, that describe when the rule is enabled to spike. Also in this case, the handbook [27] is a valuable resource to obtain further information, but the reader should also be aware that several further variants of spiking neural P systems have been defined and investigated in the last few years.

For an updated bibliography, we refer the reader to the P systems web page [31] as well as to the web page of the International Membrane Computing Society [14] that publish a Bulletin (four issues per year) containing the latest news about P systems, and promotes the International series of *Conferences on Membrane Computing* (CMC) and *Asian Conferences on Membrane Computing* (ACMC). By consulting the Bulletin and the Proceedings of these conferences, the reader will discover many other interesting P system models, such as (*enzymatic*) *numerical P systems*, *P automata*, *population P systems*, *metabolic P systems*, P systems with *antimatter*, SN P systems with *antispikes*, and still *probabilistic* and *quantum* and *asynchronous* variants, operating under different modes of applying the rules. There is an entire world to be explored!

We conclude by remarking that in this chapter, we took a *theoretical* approach, interested to determine the computational properties of P systems when considered as abstract models of computation. From the point of view of applications, there is once again a considerable literature on using P systems to model different kinds of biological and natural phenomena, and even more number of papers and software tools dealing with efficient simulations of several variants of P systems, exploiting the parallel computing power of modern GPUs. The above-mentioned references and resources provide, also in this case, the best pointers to start with.

Bibliography

[1] Aaronson, S. Why philosophers should care about computational complexity. Technical report, http://eccc.hpi-web.de/report/2011/108/, 2011.

[2] Adleman, L. M. Molecular computation of solutions to combinatorial problems. *Science* **266**(5187), 1021–1024, 1994.

[3] Barbuti, R., Maggiolo-Schettini, A., Milazzo, P., Pardini, G. and Tesei, L. Spatial P systems. *Nat. Comput.* **10**(1), 3–16, 2011.

[4] Cook, S. A. The complexity of theorem-proving procedures. In *Proceedings of the Third Annual ACM Symposium on Theory of Computing, STOC '71*, pp. 151–158, 1971.

[5] Copeland, J. (ed.). *The Essential Turing: Seminal Writings in Computing, Logic, Philosophy, Artificial Intelligence, and Artificial Life, Plus the Secrets of Enigma.* Oxford University Press, 2004.

[6] Davis, M. (ed.). *The Undecidable: Basic Papers on Undecidable Propositions, Unsolvable Problems and Computable Functions.* Raven Press, 1965.

[7] Edmonds, J. Paths, trees and flowers. *Can. J. Math.* **17**, 449–467, 1965.

[8] Fischer, P. C., Meyer, A. R. and Rosenberg, A. L. Counter machines and counter languages. *Math. Syst. Theory* **2**(3), 265–283, 1968.

[9] Fortnow, L. and Homer, S. A short history of computational complexity. *Bull. Eur. Assoc. Theor. Comput. Sci.* **80**, 95–133, 2003.

[10] Frisco, P. P systems, Petri nets, and program machines. In R. Freund, G. Păun, G. Rozenberg and A. Salomaa (eds.). *Membrane Computing, 6th International Workshop, WMC 2005*, Lecture Notes in Computer Science, Vol. 3850, pp. 209–223. Springer, 2006.

[11] Frisco, P. *Computing with Cells: Advances in Membrane Computing.* Oxford University Press, New York, 2009.

[12] Hartmanis, J. and Stearns, R. E. On the computational complexity of algorithms. *Trans. Amer. Math. Soc.* **117**, 285–306, 1965.

[13] Head, T. Formal language theory and DNA: An analysis of the generative capacity of specific recombinant behaviors. *Bull. Math. Biol.* **49**(6), 737–759, 1987.

[14] International Membrane Computing Society (IMCS). http://membranecomputing.net/.

[15] Ionescu, M., Păun, G. and Yokomori, T. Spiking neural P systems. *Fundam. Inform.* **71**(2, 3), 279–308, 2006.

[16] Karp, R. M. Reducibility among combinatorial problems. In *Complexity of Computer Computations*, pp. 85–104. Plenum Press, 1972.

[17] Kuroiwa, T., Kuroiwa, H., Sakai, A., Takahashi, H., Toda, K. and Itoh, R. The division apparatus of plastids and mitochondria. *Int. Rev. Cytol.* **181**, 1–41, 1998.

[18] Leporati, A., Manzoni, L., Mauri, G., Porreca, A. E. and Zandron, C. Characterising the complexity of tissue P systems with fission rules. *J. Comput. Syst. Sci.* **90**, 115–128, 2017.

[19] Leporati, A., Manzoni, L., Mauri, G., Porreca, A. E. and Zandron, C. A toolbox for simpler active membrane algorithms. *Theoret. Comput. Sci.* **673**, 42–57, 2017.
[20] Martín-Vide, C., Păun, G., Pazos, J. and Rodríguez-Patón, A. Tissue P systems. *Theoret. Comput. Sci.* **296**(2), 295–326, 2003.
[21] McCulloch, W. S. and Pitts, W. A logical calculus of the ideas immanent in nervous activity. *Bull. Math. Biophys.* **7**, 115–133, 1943.
[22] Minsky, M. *Computation: Finite and Infinite Machines.* Prentice-Hall, 1967.
[23] Papadimitriou, C. H. *Computational Complexity.* Addison-Wesley, 1993.
[24] Păun, G. Computing with membranes. *J. Comput. Syst. Sci.* **61**(1), 108–143, 2000.
[25] Păun, G. P systems with active membranes: Attacking NP-complete problems. *J. Autom. Lang. Combin.* **6**(1), 75–90, 2001.
[26] Păun, G. Membrane computing: History and brief introduction. In E. Gelenbe and J.-P. Kahane (eds.). *Fundamental Concepts in Computer Science*, pp. 17–41. Imperial College Press, 2009.
[27] Păun, G., Rozenberg, G. and Salomaa, A. (eds.) *The Oxford Handbook of Membrane Computing.* Oxford University Press, 2010.
[28] Păun, A. and Păun, G. The power of communication: P systems with symport/antiport. *New Gener. Comput.* **20**(3), 295–305, 2002.
[29] Păun, G., Pérez-Jiménez, M. J. and Riscos-Núñez, A. Tissue P systems with cell division. *Int. J. Comput. Commun. Control* **3**(3), 295–303, 2008.
[30] Sosík, P. and Rodríguez-Patón, A. Membrane computing and complexity theory: A characterization of PSPACE. *J. Comput. Syst. Sci.* **73**(1), 137–152, 2007.
[31] P systems web site. http://ppage.psystems.eu/.
[32] Turing, A. M. On computable numbers, with an application to the Entscheidungsproblem. *Proc. London Math. Soc.* **42**, 230–265, 1936.
[33] Vollmer, H. *Introduction to Circuit Complexity: A Uniform Approach.* Texts in Theoretical Computer Science: An EATCS Series. Springer, 1999.
[34] Zach, R. Hilbert's program then an now. In D. Jacquette (ed.). *Philosophy of Logic*, Vol. 5, pp. 411–447. Elsevier, 2006.

Chapter 2

Results on Computational Complexity in Bio-inspired Computing

Mario J. Pérez-Jiménez*, Agustín Riscos-Núñez,
Luis Valencia-Cabrera and David Orellana-Martín

Research Group on Natural Computing
Department of Computer Science and
Artificial Intelligence, Universidad de Sevilla
Avda. Reina Mercedes s/n
41012 Sevilla, Spain
**marper@us.es*

It is well known that the calculation speed of electronic processors has an upper bound which is clearly insufficient to efficiently solve instances whose size is relatively large associated with problems which model relevant real-life problems. In the last few decades, several computational paradigms inspired from nature have been developed with the aim of overcoming the barrier previously mentioned by using devices of the new computing models. For this purpose, the space–time trade-off method is used to develop efficient solutions to computationally hard problems. According to this, implementation of such models (in biological or any other substrate different from the electronic one) would provide a significant advance in the practical resolution of many real-life problems. Membrane computing is a computational paradigm motivated by the structure and functioning of the living cells, as well as from the organization of cells in tissues, organs, and other higher order structures. In this chapter, a computational complexity theory within the framework of Membrane Computing is introduced and the most relevant results obtained so far are presented.

2.1. Introduction

Computability Theory deals with "mechanical" tasks and studies those procedures ("algorithms", in a generic way) than can be implemented by these tasks. The theory tries to define the informal idea of algorithmic problems resolution in a rigorous way. Each formal definition of the said concept provides a computing model. A basic question is to determine the class of all the problems that can be solved by a computing model when using the procedures/algorithms defined in it. In any computing model which captures the informal idea of the algorithm, there are undecidable problems, that is, problems that cannot be solved by using the algorithms of the model.

Analyzing an algorithm which solves a problem consists of determining the amount of computational resources required to perform it. One of the main goals of *Complexity Theory* is to study the minimal resource requirements with which the problem can be solved, and to provide tools that allow the classification of abstract problems regarding the amount of resources needed for their resolution (that is, according to their intrinsic complexity of computational tasks). These classes allow us to detect some of the inherent difficulties in the computational resolution of some problems.

The **P** versus **NP** question is the problem of determining whether every problem solvable by means of some non-deterministic Turing machine in polynomial time can also be solved by some deterministic Turing machine in polynomial time. In other words, this question can be informally reformulated as asking whether or not finding solutions of a (search) problem is harder than checking the correctness of a given solution of it. It can also be expressed as asking whether or not deciding membership in a set is harder than being convinced of this membership by an adequate proof [7]. The classical approach when trying to solve this question is to focus on one **NP**-complete problem and try to prove that it belongs (or not) to the class **P**. A membership proof can be obtained by designing a deterministic algorithm solving that problem in polynomial time, and in such case, we immediately deduce that **P** = **NP** holds. In the second case, claiming that the problem does not belong to **P** is equivalent to proving that no deterministic algorithm working in polynomial time provides a solution to the problem, and therefore in this case we deduce $\mathbf{P} \neq \mathbf{NP}$.

The **P** versus **NP** question is one of the outstanding open problems in theoretical computer science. A negative answer to this question would confirm that the majority of current cryptographic systems are secure from a practical point of view. A positive answer would not only show the uncertainty about the security of these systems, but also this kind of answer is expected to come together with a general procedure that provides a deterministic algorithm solving most of the **NP**-complete problems in polynomial time [34].

2.2. Preliminaries

In order to make this chapter self-contained, some basic concepts and results are presented in this section.

2.2.1. *Multisets over an alphabet*

An *alphabet* Γ is a non-empty set and their elements are called *symbols*. A *string* u over Γ is an ordered finite sequence of symbols, that is, a mapping from a natural number $n \in \mathbb{N}$ onto Γ, where number n is called the *length* of the string u and it is denoted by $|u|$. The empty string (with length 0) is denoted by λ. The set of all strings over an alphabet Γ is denoted by Γ^*. A *language* over Γ is a subset of Γ^*.

A *multiset* over an alphabet Γ is an ordered pair (Γ, f), where f is a mapping from Γ onto the set of natural numbers $\mathbb{N}$. The *support* of a multiset $m = (\Gamma, f)$ is defined as $\text{supp}(m) = \{x \in \Gamma \mid f(x) > 0\}$. A multiset is finite (respectively, empty) if its support is a finite (respectively, empty) set. We denote the empty multiset by $\emptyset$.

Let $m_1 = (\Gamma, f_1)$ and $m_2 = (\Gamma, f_2)$ be multisets over Γ, then the union of m_1 and m_2, denoted by $m_1 + m_2$, is the multiset (Γ, g), where $g(x) = f_1(x) + f_2(x)$ for each $x \in \Gamma$. We say that m_1 is contained in m_2 and we denote it by $m_1 \subseteq m_2$, if $f_1(x) \leq f_2(x)$ for each $x \in \Gamma$.

2.2.2. *Graphs and trees*

Let us recall some notions related to the graph theory (see [4] for details). An *undirected graph* is an ordered pair (V, E), where V is a set whose elements are called nodes or vertices and $E = \{x, y\} \mid x \in V, y \in V, x \neq y\}$ whose elements are called *edges*. A *directed graph* is an ordered pair (V, E), where V is a set whose elements are called nodes or vertices and $E = \{(x, y) \mid x \in V, y \in V\}$ whose elements are called *arcs*. A *path* of length $k \geq 1$ from a node u to a node v in a graph (V, E) is a finite sequence $(x_0, x_1, \ldots, x_k)$ of nodes such that $x_0 = u$, $x_k = v$ and $(x_i, x_{i+1}) \in E$ for $0 \leq i \leq k-1$ (in the case of a directed graph) or $\{x_i, x_{i+1}\} \in E$ (in the case of an undirected graph). If $k \geq 2$ and $x_0 = x_k$, then we say that the path is a *cycle* of the graph. A graph with no cycles is said to be *acyclic*. An undirected graph is *connected* if there exist paths between every pair of nodes.

A *free tree* (*tree*, for short) is a connected, acyclic, undirected graph. A *rooted tree* is a tree in which one of the vertices (called *the root of the tree*) is distinguished from the others. In a rooted tree, the concepts of ascendants and descendants are defined in a usual way. Given a node x (different from the root) in a rooted tree, if the last edge on the (unique) path from the root to the node x is $\{x, y\}$ (so $x \neq y$), then y is **the** *parent* of node x and x is **a** *child* of node y. We denote it by $y = p(x)$ and $x \in ch(y)$. The root is the only node in the tree with no parent. A node with no children is called a *leaf*.

Let us recall the `Reachability Problem`: *given a (directed or undirected) graph, G, and two nodes a, b, determine whether or not the node b is reachable from a, that is, whether or not there exists a path in the graph from a to b.* We denote by `Reachability`(G, a, b) the answer (`yes` or `no`) to the Reachability problem with instance (G, a, b). It is easy to design an algorithm running in polynomial time solving this problem. For example, given a (directed or undirected) graph, G, and two nodes a, b, we consider a depth-first search with source a, and we check if b is in the tree of the computation forest whose root is a. The total running time of this algorithm is $O(|V| + |E|)$, that is, in the worst case is quadratic in the number of nodes.

Moreover, this algorithm needs to store a linear number of items (it can be proved that there exists another polynomial time algorithm which uses $O(\log^2(|V|))$ space.

2.2.3. *Decision problems and languages*

Roughly speaking, a decision problem X is one whose solution/ answer is either "yes" or "no". This can be formally defined by an ordered pair (I_X, θ_X), where I_X is a language over a finite alphabet Σ_X and θ_X is a total Boolean function over I_X. The elements of I_X are called *instances* of the problem X. Each decision problem X has a language L_X over the alphabet Σ_X associated as follows: $L_X = \{u \in E_X \mid \theta_X(u) = 1\}$. Conversely, every language L over an alphabet Σ has a decision problem $X_L = (I_{X_L}, \theta_{X_L})$ associated as follows: $I_{X_L} = \Sigma^*$ and $\theta_{X_L}(u) = 1$ if and only if $u \in L$. Then, given a decision problem X, we have $X_{L_X} = X$, and given a language L over an alphabet Σ, we have $L_{X_L} = L$.

The SAT (*Propositional Satisfiability*) problem is the following decision problem: *given a simplified Boolean formula in conjunctive normal form, determine whether or not there exists an assignment of truth values to its variables on which that formula evaluates true.*

The HAM-CYCLE (*Hamiltonian Cycle*) problem is the following decision problem: *given a directed graph, determine whether or not there exists a Hamiltonian cycle (i.e. a cycle that visits each vertex exactly once) in the graph.*

2.2.4. *The Cantor pairing function*

The Cantor pairing function encodes pairs of natural numbers by single natural numbers and it is defined as follows. For each $n, p \in \mathbb{N}$

$$\langle n, p \rangle = \frac{(n+p)(n+p+1)}{2} + n$$

The Cantor pairing function is a primitive recursive function and bijective from $\mathbb{N} \times \mathbb{N}$ onto $\mathbb{N}$. Then, for each $t \in \mathbb{N}$, there exist unique natural numbers $n, p \in \mathbb{N}$ such that $t = \langle n, p \rangle$.

2.3. Membrane Systems

Membrane Computing is a branch of *Natural Computing* initiated by Gh. Păun at the end of 1998. It is inspired by some basic biological features and by the way the living cells process chemical substances, energy and information. The devices of this computing paradigm, called *P systems* or *membrane systems*, constitute models for distributed, parallel and non-deterministic computing.

In Membrane Computing, there are basically two ways to consider computational devices: cell-like membrane systems (*P systems*) and tissue-like membrane systems (*tissue P systems*). The first ones use biological membranes arranged hierarchically and are inspired by the structure of the cell, while the second ones use biological cells placed in the nodes of a graph, inspired by the cell intercommunication in tissues. Both of them have an associated *environment* which represents the region surrounding the system.

A *membrane system* can be viewed as a finite set of basic units of processors arranged in a structure (a rooted tree in the case of a cell-like approach and a directed graph in the case of a tissue-like approach) and delimiting *regions* able to contain multisets of *objects* (elements of a *working alphabet* Γ) which are abstractions of chemical substances, and finite sets of *rewriting rules* which are abstractions of chemical reactions, that provide the dynamics of the system. The objects that evolve in a transition step and the rules by which they evolve are chosen in a non-deterministic way and, in general, in a maximally parallel manner: we assign objects to rules but in such a way that after this assignation no further rule can be applied to the remaining objects. There exists a double parallelism in the application of rules, once at the level of each region/processor (the rules are used in parallel) and once at the level of the system (all regions/processors evolve concomitantly).

For cell-like systems, a rooted tree structure is explicitly defined: its nodes are called *membranes*, the root of the tree is called the *skin membrane* of the system, and the leaves of the tree are called *elementary membranes*. In this kind of membrane systems, the environment plays a passive role in the sense that it is initially empty, and it only receives objects from the system through the root of the tree (the skin membrane) but it cannot send objects inside the system. For each membrane i (different from the skin membrane) we denote by $p(i)$, the parent of membrane i in the rooted tree and, by convention, the "parent" of the skin membrane is the environment.

In tissue-like systems, the graph structure is implicitly defined through some especial rules of the system (*communication rules*), and its nodes are called *cells*. In this kind of membrane systems, the environment plays an active role since it may interact with any cell of the system; not only receiving objects, but also sending objects inside the cells. Besides, the environment has initially an arbitrary large number of copies of objects from a distinguished alphabet, $\mathcal{E}$, strictly contained within the working alphabet.

With respect to the semantics ingredients, a *configuration* at any instant of a membrane system is described by all multisets of objects over the working alphabet associated with all the processors (membranes or cells) present in the system together with the current tree structure. In tissue-like systems, we also consider the multiset of objects from $\Gamma \setminus \mathcal{E}$ present in the environment at that moment. We can pass from a configuration $\mathcal{C}_1$ to a configuration $\mathcal{C}_2$ in one *transition step* if $\mathcal{C}_2$ is obtained from $\mathcal{C}_1$ by applying the rules from the system. A *computation* is a (finite or infinite) sequence of configurations such that: (a) the first term of the sequence is the initial configuration of the system; (b) each remaining term is obtained from the previous configuration by applying rules of the system; and (c) if the sequence is finite (called *halting computation*), then the last term of the sequence must be a halting configuration.

Throughout this chapter any membrane system has a distinguished alphabet, the *input alphabet*, Σ, strictly contained in the working alphabet, Γ, which allows us to encode instances of a problem and these codes are introduced in the system by using a distinguished (input) region/processor. The initial multisets of the system associated with regions are multisets over $\Gamma \setminus \Sigma$. Given a membrane system Π, for each multiset m over the input alphabet we have an initial configuration associated with it, which is obtained by adding m to the initial multiset associated with the input region. It will be denoted by $\Pi + m$.

2.3.1. *Recognizer membrane systems*

In order to study the computational efficiency of membrane systems, that is, the ability to solve computationally hard problems in an efficient way, the notions from classical *computational complexity theory* are adapted for Membrane Computing. In this context, a special class of cell-like P systems was introduced in [32]: *recognizer P systems* (called *accepting P systems* in a previous paper [27]). For tissue P systems, with the same idea as recognizer cell-like P systems, *recognizer tissue P systems* were introduced in [24].

A membrane system is a *recognizer* system if: (a) the working alphabet has two distinguished objects yes and no, but in the case of tissue P systems, at least one copy of them is present in some initial multisets but none of them are present in the environment initially; (b) all computations halt; and (c) for each computation of the system, either object yes or object no (but not both) must have been released into the environment, and only at the last step of the computation.

We say that a computation $\mathcal{C}$ of a recognizer membrane system is an *accepting computation* (respectively, *rejecting computation*) if object yes (respectively, object no) appears in the environment associated with the corresponding halting configuration of $\mathcal{C}$, and neither object yes nor no appears in the environment associated with any non-halting configuration of $\mathcal{C}$.

2.3.2. *Polynomial complexity classes of membrane systems*

Let us recall that a *decision problem* is a pair (I_X, θ_X), where I_X is a language over a finite alphabet (whose elements are called *instances*) and θ_X is a total Boolean function over I_X. Next, we define what means solving a decision problem in the framework of membrane systems efficiently and in a uniform way.

Definition 2.1. We say that a decision problem $X = (I_X, \theta_X)$ is solvable in a uniform way and in polynomial time by a family $\mathbf{\Pi} = \{\Pi(n) \mid n \in \mathbb{N}\}$ of recognizer membrane systems from a class $\mathcal{R}$ (we denote it by $X \in \mathbf{PMC}_{\mathcal{R}}$) if the following conditions hold:

(1) Family $\mathbf{\Pi}$ is *polynomially uniform* by Turing machines, that is, there exists a deterministic Turing machine working in polynomial time such that on input 1^n, system $\Pi(n)$ is constructed.

(2) There exists a pair (*cod*, s) of polynomial-time computable functions over I_X such that:

  (a) for each instance $u \in I_X$, $s(u)$ is a natural number and $cod(u)$ is an input multiset of the system $\Pi(s(u))$,
  (b) for each $n \in \mathbb{N}$, $s^{-1}(n)$ is a finite set,
  (c) family $\mathbf{\Pi}$ is *polynomially bounded* with regard to (X, cod, s), that is, there exists a polynomial function p such that for each $u \in I_X$, every computation of $\Pi(s(u))$ with input $cod(u)$ is halting and it performs at most $p(|u|)$ steps,
  (d) family $\mathbf{\Pi}$ is *sound* with regard to (X, cod, s), that is, for each $u \in I_X$, if **there exists** an accepting computation of $\Pi(s(u))$ with input $cod(u)$, then $\theta_X(u) = 1$,
  (e) family $\mathbf{\Pi}$ is *complete* with regard to (X, cod, s), that is, for each $u \in I_X$, if $\theta_X(u) = 1$, then **every** computation of $\Pi(s(u))$ with input $cod(u)$ is an accepting one.

If $\mathcal{R}$ is a class of recognizer membrane systems, then we denote by $\mathbf{PMC}_{\mathcal{R}}$ the set of all decision problems which can be solved in a uniform way and in polynomial time by means of families of systems from $\mathcal{R}$. Class $\mathbf{PMC}_{\mathcal{R}}$ is closed under complement and polynomial-time reductions [27].

Let us note that from the soundness and completeness conditions above, we deduce that family $\mathbf{\Pi}$ is *confluent* with regard to (X, cod, s), that is, for each $u \in I_X$, either all the computations of system $\Pi(s(u)) + cod(u)$ are accepting computations, or all of them are rejecting ones.

It is worth pointing out that there is an important difference between the notion of acceptance by a non-deterministic Turing machine or by means of a recognizer membrane system. Let us recall that an instance of a decision problem is accepted by a non-deterministic Turing machine processing it, if and only if there exists, at least, an accepting computation. Nevertheless, an instance will be accepted by a membrane system processing it, if and only if each computation of the system is an accepting computation.

2.4. Frontiers of the Efficiency in Membrane Systems

In this section, the concepts of *efficiency* and *feasibility* in membrane systems are introduced and we study different classes of recognizer membrane systems with respect to some syntactical ingredients.

Definition 2.2. We say that a class of recognizer membrane systems $\mathcal{R}$ is **presumably efficient** if there exists an **NP**-complete problem that can be solved in a uniform way and in polynomial time by a family of membrane systems from $\mathcal{R}$.

From the properties of **NP**-completeness, we deduce that *any* **NP**-complete problem can be solved in a uniform way and in polynomial time by families of a presumably efficient class of recognizer membrane systems. In other words, a class of recognizer membrane systems $\mathcal{R}$ is presumably efficient if and only if $\mathbf{NP} \cup \mathbf{co\text{-}NP} \subseteq \mathbf{PMC}_{\mathcal{R}}$.

Definition 2.3. We say that a class of recognizer membrane systems $\mathcal{R}$ is **feasible** if only problems in class **P** can be solved in a uniform way and in polynomial time by a family of membrane systems from $\mathcal{R}$, that is, if and only if $\mathbf{PMC}_{\mathcal{R}} = \mathbf{P}$.

According to these definitions, if $\mathbf{P} \neq \mathbf{NP}$, then the term "feasible" is equivalent to the term "non-presumably efficient", in particular, each feasible class is not presumably efficient. Let us recall that, in this case, as a consequence of the Ladner theorem, there exist **NP**-intermediate problems, that is, problems which are neither in class **P** nor in class of **NP**-complete problems (see [12] for details). On the other hand, if **P** = **NP**, then a class $\mathcal{R}$ is feasible if and only if it is presumably efficient.

Let us suppose we have two classes $\mathcal{R}_1$ and $\mathcal{R}_2$ of recognizer membrane systems such that:

(a) $\mathcal{R}_1 \subsetneq \mathcal{R}_2$ in the following sense: membrane systems from $\mathcal{R}_2$ are obtained from membrane systems from $\mathcal{R}_1$ by adding some syntactical (and/or semantics) ingredients. Then, each solution of a decision problem X in $\mathcal{R}_1$ is also a solution of X in $\mathcal{R}_2$;
(b) $\mathcal{R}_1$ is feasible, that is, $\mathbf{PMC}_{\mathcal{R}_1} = \mathbf{P}$;
(c) $\mathcal{R}_2$ is presumably efficient, that is, **NP**-complete problems can be solved in a uniform way and in polynomial time in $\mathcal{R}_2$.

Then, the syntactical ingredients which are added to membrane systems in $\mathcal{R}_1$ to obtain membrane systems in $\mathcal{R}_2$ provide a borderline of efficiency. In this context, for instance, a strategy to prove that **P** = **NP** would be the following: to *translate* an efficient solution of an **NP**-complete problem by a family of systems in $\mathcal{R}_2$ into an efficient solution by a family of systems in $\mathcal{R}_1$.

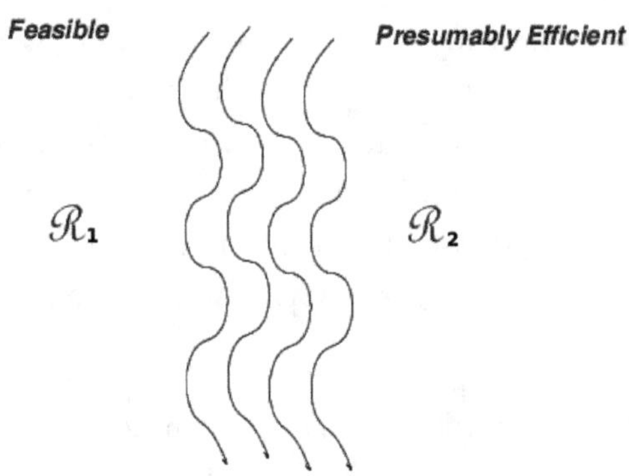

In order to state the feasibility or the presumably efficiency of a class $\mathcal{R}$ of recognizer membrane systems, three important techniques have been basically used: *dependency graph technique*, *simulation technique* and *algorithmic technique*.

2.4.1. *Dependency graph technique*

By means of this technique, a directed graph is associated with each membrane system Π where each rule can be activated by a single object. In this case, there exists in some sense, a *dependency* between the object triggering the rule and the object or objects produced by its application. In this situation, a directed graph (*dependency graph*) will be associated with Π verifying the following relevant property: there exists an accepting computation of Π if and only if there exists a path between two distinguished nodes in the dependency graph associated with it (see [10, 11] for more details).

In order to apply the dependency graph technique, it is worth pointing out that the `Reachability Problem` (*given a — directed or undirected — graph and two nodes* a, b, *determine whether or not there exists a path in the graph from node* a *to node* b) belongs to the complexity class **P**.

2.4.2. *Simulation technique*

The concept of *efficient simulations* in the framework of recognizer membrane systems is introduced as follows. Given two classes $\mathcal{R}$ and $\mathcal{R}'$ of recognizer membrane systems we say that $\mathcal{R}$ *is simulated* by $\mathcal{R}'$ *in an efficient way* if there exists a deterministic Turing machine working in polynomial time such that for each membrane system Π from $\mathcal{R}$, a membrane system Π' from $\mathcal{R}'$ is provided in such a manner that there exists an injective function, f, from the set $\mathbf{Comp}(\Pi)$ of computations of Π onto the set $\mathbf{Comp}(\Pi')$ of computations of Π' such that:

- There exists a deterministic Turing machine that constructs computation $f(\mathcal{C})$ from computation $\mathcal{C}$ in polynomial time.
- A computation $\mathcal{C} \in \mathbf{Comp}(\Pi)$ is an accepting computation if and only if $f(\mathcal{C}) \in \mathbf{Comp}(\Pi')$ is an accepting one.
- There exists a polynomial function $p(n)$ such that for each $\mathcal{C} \in \mathbf{Comp}(\Pi)$ we have $|f(\mathcal{C})| \leq p(|\mathcal{C}|)$.

This technique can be used when we try to prove that polynomial complexity class $\mathbf{PMC}_{\mathcal{R}}$ is contained in class $\mathbf{PMC}_{\mathcal{R}'}$.

2.4.3. *Algorithmic technique*

This technique can be used to prove that only problems in class **P** can be solved in polynomial time by families of systems from a class $\mathcal{R}$ of recognizer membrane systems. The technique consists in constructing a deterministic algorithm $\mathcal{A}$ working in polynomial time that receives as input a recognizer membrane system Π from $\mathcal{R}$ and an input multiset m of Π. Then, algorithm $\mathcal{A}$ reproduces the behaviour of a computation of $\Pi + m$. In particular, if the given recognizer membrane system is confluent, then the algorithm will provide the same answer of the system, that is, the answer of algorithm $\mathcal{A}$ is affirmative if and only if system $\Pi + m$ has an accepting computation (and then, any computation is an accepting one).

2.5. Efficiency of Cell-Like Membrane Systems

In this section, the computational efficiency of several models of cell-like membrane systems is studied. These variants are considered by using a specific kind of rules, and by adding new ingredients as electrical charges too.

2.5.1. *Basic transition P systems*

Basic transition P systems are a kind of cell-like membrane systems whose membrane structure does not grow, that is, the number of membranes can only decrease during a computation, by dissolving membranes as a result of applying evolution rules to the objects present in the system.

Definition 2.4. A recognizer basic transition P system of degree $q \geq 1$ is a tuple of the form $\Pi = (\Gamma, \mu, \mathcal{M}_1, \dots, \mathcal{M}_q, \mathcal{R}_1, \dots, \mathcal{R}_q, i_{in})$, where:

- Γ is a finite alphabet,
- μ is a rooted tree whose nodes are injectively labelled with $1, \dots, q$,
- $\mathcal{M}_1, \dots, \mathcal{M}_q$ are finite multisets over Γ,

- $\mathcal{R}_i$, $1 \le i \le q$, are finite sets of rules over Γ of the form $u \to v$, where $v = (v_1, \text{here})\ (v_2, \text{out})\ (v_3, in_j)\delta^*$, being u, v_1, v_2, v_3 finite multisets over Γ, $1 \le j \le q$ and $\delta^* \in \{\lambda, \delta\}$,
- $i_{in} \in \{1, \dots, q\}$.

The class of all recognizer basic transition P systems will be denoted by $\mathcal{T}$.

A *configuration* $\mathcal{C}_t$ at an instant t of a P system is a tuple whose components are the membrane structure at instant t and all multisets of objects over Γ associated with all the membranes present in μ. The *initial configuration* of Π is $(\mu, \mathcal{M}_1, \dots, \mathcal{M}_q)$. A rule of the form $u \to (v_1, here)(v_2, out)(v_3, in_j)\,\delta^* \in \mathcal{R}_i$ is *applicable* to a configuration $\mathcal{C}_t$ at an instant t if the following conditions hold: (a) membrane i is in $\mathcal{C}_t$; (b) multiset u is contained in the multiset associated with such membrane; (c) if multiset v_3 is non-empty, then $p(j) = i$; and (d) if $\delta^* = \delta$, then membrane i is different from the skin membrane ($i \neq 1$) and different from the output membrane, if any. When applying such a rule, all objects in v_1 will be placed in the same region i where the rule is applied; all objects in v_2 will be placed in the (parent) region $p(i)$, all objects in v_3 will be placed in the region j, and if $\delta^* = \delta$, then membrane i is dissolved. After dissolving a membrane, all objects and membranes previously present in it become elements of the contents of the immediately upper membrane that has not been dissolved. A *transition step* from a configuration $\mathcal{C}_1$ to a configuration $\mathcal{C}_2$ is obtained by applying the rules in a maximally parallel manner.

First of all, in order to formally define what that a family of membrane systems simulates a Turing machine means, we shall introduce for each Turing machine a decision problem associated with it.

Definition 2.5. Let M be a Turing machine with input alphabet Σ_M. The decision problem associated with M is the problem $X_M = (I_M, \theta_M)$, where $I_M = \Sigma_M^*$, and for every $w \in \Sigma_M^*$, $\theta_M(w) = 1$ if and only if M accepts w.

Obviously, the decision problem X_M is solvable by the Turing machine M.

Definition 2.6. We say that a Turing machine M is simulated in polynomial time by a family of recognizer membrane systems from $\mathcal{R}$ if and only $X_M \in \mathbf{PMC}_{\mathcal{R}}$.

In [11], an efficient simulation of deterministic Turing machines by recognizer basic transition P systems was given.

Proposition 2.1 (Sevilla theorem). *Every deterministic Turing machine working in polynomial time can be simulated in polynomial time by a family of recognizer basic transition P systems.*

In [11], it was also proved that each confluent basic transition P system (every computation produces the same answer) can be efficiently simulated by a deterministic Turing machine [11]. As a consequence, only problems in class **P** can be solved in polynomial time by means of families of this kind of membrane systems.

Proposition 2.2. *If a decision problem is solvable in polynomial time by a family of recognizer basic transition P systems, then there exists a deterministic Turing machine solving it in polynomial time.*

Therefore, the following theorem holds.

Theorem 2.1. $\mathbf{P} = \mathbf{PMC}_{\mathcal{T}}$.

Thus, assuming that $\mathbf{P} \neq \mathbf{NP}$, the ability of cell-like membrane systems in $\mathcal{T}$ to create exponential workspace (in terms of number of objects) in polynomial time (e.g. via evolution rules of the type $[a \to a^2]_h$) is not enough to efficiently solve **NP**-complete problems.

Corollary 2.1. $\mathbf{P} \neq \mathbf{NP}$ *if and only if every* (*or at least one*) **NP**-*complete problem is not in* $\mathbf{PMC}_{\mathcal{T}}$.

This result provides a tool to attack conjecture $\mathbf{P} \neq \mathbf{NP}$ in the framework of Membrane Computing.

2.5.2. *P systems with active membranes*

Cell division is one of the most important processes of a cell that produces two or more cells from one cell by *replicating* the original chromosomes so that each daughter cell gets a copy of it. Mitosis is an elegant process of cell division which results in the production of two daughter cells from a single parent cell. Daughter cells are identical to each other and to the original parent cell. Through a sequence of steps, the replicated genetic material in a parent cell is equally distributed to two daughter cells. While there are some subtle differences, mitosis is remarkably similar across organisms. This mechanism was introduced in Membrane Computing through rules called *membrane division* that are triggered by an object which is replaced in the two new cells by possibly new objects and the remaining objects are *duplicated* in both new cells/membranes. This has given rise to *P systems with active membranes* [21]. It is a kind of cell-like membrane system having electrical charges associated with membranes, and they were first introduced by Gh. Păun in [21]. These systems incorporate membrane division rules, do not use cooperation, and the polarization of a membrane (but not the label) can be modified by the application of a rule.

The rules of a P system with active membranes are of the following forms:

(a) $[a \to u]_h^{\alpha}$ (*object evolution rules*).
(b) $a[\,]_h^{\alpha_1} \to [b]_h^{\alpha_2}$ (*send-in communication rules*).
(c) $[a]_h^{\alpha_1} \to [\ ]_h^{\alpha_2}\, b$ (*send-out communication rules*).
(d) $[a]_h^{\alpha} \to b$ (*dissolution rules*).
(e) $[a]_h^{\alpha_1} \to [b]_h^{\alpha_2}\,[c]_h^{\alpha_3}$ (*division rules for elementary membranes*).
(f) $[[\,]_{h_1}^{\alpha_1}\,[\,]_{h_2}^{\alpha_2}]_h^{\alpha} \to [[\,]_{h_1}^{\alpha_3}]_h^{\beta}\,[[\,]_{h_2}^{\alpha_4}]_h^{\gamma}$ (*division rules for non-elementary membranes*).

Here a, b, c are objects, u is a multiset of objects and α_i are electrical charges from the set $\{+, 0, -\}$.

The rules of a P system with active membranes are applied according to the following principles (see [22] for details):

- In one step, one object of a membrane can be used by only one rule (chosen in a non-deterministic way).
- If a membrane is dissolved, its content (multiset and internal membranes) is left free in the surrounding region.
- If at the same time a membrane labelled by h is divided by a rule of type (e) and there are objects in this membrane which can evolve by means of rules of type (a), then "first" each object evolution rule is applied, and then division takes place. Of course, this process takes only one step. The skin membrane cannot be divided.
- The rules associated with membranes labelled by h are used for all copies of this membrane. At one step, a membrane can be the subject of *only one* rule of types (b)–(f).

Let us denote by $\mathcal{AM}$ the class of recognizer P systems with active membranes, and let $\mathcal{NAM}$ be the class of recognizer P systems with active membranes which do not make use of division rules.

In the framework of cell-like membrane systems, confluent recognizer P systems with active membranes making use of no membrane division rule can be efficiently simulated by a deterministic Turing machine [41].

Proposition 2.3 (Milano theorem). *A deterministic P system with active membranes but without membrane division can be simulated by a deterministic Turing machine with a polynomial slowdown.*

As a consequence of the previous result, the following corollary holds.

Corollary 2.2. $\mathbf{PMC}_{\mathcal{NAM}} \subseteq \mathbf{P}$.

In [37], a simple proof showing that each tractable problem can be solved by a family of recognizer P systems with active membranes (without polarizations) operating in exactly one step and using only send-out communication rules has been provided.

Proposition 2.4. $\mathbf{P} \subseteq \mathbf{PMC}_{\mathcal{NAM}}$.

Thus, we have a version of Theorem 2.1 for the class $\mathcal{NAM}$.

Theorem 2.2. $\mathbf{P} = \mathbf{PMC}_{\mathcal{NAM}}$.

Remark 2.1. In the framework of P systems with active membranes, the use or not of the division rules provides a borderline for the tractability of decision problems, assuming that $\mathbf{P} \neq \mathbf{NP}$, that is, by using division rules, we can solve **NP**-complete problems in polynomial time, but without division rules only problems in **P** can be solved in an efficient way.

We denote by $\mathcal{AM}(+n)$ (respectively, $\mathcal{AM}(-n)$) the class of recognizer P systems with active membranes using division rules for elementary and non-elementary membranes (respectively, only for elementary membranes).

In the framework of $\mathcal{AM}(-n)$, efficient solutions to weakly **NP**-complete problems (`Knapsack` [28], `Subset Sum` [29], `Partition` [9]) and strongly **NP**-complete problems (`SAT` [27], `Clique` [2], `Bin Packing` [30], `Common Algorithmic Problem` [31]) have been given.

Proposition 2.5. $\mathtt{SAT} \in \mathbf{PMC}_{\mathcal{AM}(-n)}$.

Next, we describe the solution proposed in [27]. For each $m, n \in \mathbb{N}$, membrane system $\Pi(\langle m, n\rangle)$ is defined as follows:

- The input alphabet is $\Sigma(m,n) = \{x_{i,j}, \overline{x}_{i,j} \mid 1 \leq i \leq m, 1 \leq j \leq n\}$.
- The working alphabet is $\Sigma(m,n) \cup \{c_k \mid 1 \leq k \leq m+2\} \cup \{d_k \mid 1 \leq k \leq 3n+2m+3\} \cup \{r_{i,k} \mid 0 \leq i \leq m, 1 \leq k \leq 2n\} \cup \{e,t\} \cup \{Yes, No\}$.
- The set of labels is $\{1,2\}$.
- The initial membrane structure is $[[\,]_2]_1$ (we call internal membrane to each membrane with label 2).
- The initial multisets associated with the membranes are $\mathcal{M}_1 = \emptyset$ and $\mathcal{M}_2 = \{d_1\}$. The input membrane is labelled by 2.
- The set of rules is the following:
  (a) $\{[d_k]_2^0 \rightarrow [d_k]_2^+[d_k]_2^- \mid 1 \leq k \leq n\}$.
  By using these rules, an internal membrane is divided into two membranes with opposite polarizations. These rules allow us to duplicate, in one step, the total number of internal membranes.

(b) $\{[x_{i,1} \to r_{i,1}]_2^+,\ [\overline{x}_{i,1} \to r_{i,1}]_2^- \mid 1 \le i \le m\}$.
$\{[x_{i,1} \to \lambda]_2^-,\ [\overline{x}_{i,1} \to \lambda]_2^+ \mid 1 \le i \le m\}$.
These rules try to implement a process allowing the internal membranes to encode the *assignment* of a variable and, simultaneously, to check the value of all clauses by this assignment, in such a way that if the clause is true then an object $r_{i,1}$ will appear in the membrane. In other case, the object encoding the variable will disappear.

(c) $\{[x_{i,j} \to x_{i,j-1}]_2^+,\ [x_{i,j} \to x_{i,j-1}]_2^- \mid 1 \le i \le m,\ 2 \le j \le n\}$;
$\{[\overline{x}_{i,j} \to \overline{x}_{i,j-1}]_2^+,\ [\overline{x}_{i,j} \to \overline{x}_{i,j-1}]_2^- \mid 1 \le i \le m,\ 2 \le j \le n\}$.
The check process described previously is always made with respect to the *first* variable appearing in the internal membrane. Hence, the rules of (c) take charge of making a cyclic path through all the variables to get that, initially, the first variable is x_1, then x_2, and so on.

(d) $\{[d_k]_2^+ \to [\,]_2^0 d_k,\ [d_k]_2^- \to [\,]_2^0 d_k \mid 1 \le k \le n\}$;
$\{d_k[\,]_2^0 \to [d_{k+1}]_2^0 \mid 1 \le k \le n-1\}$.
These rules are used as controllers of the generating process of the assignments and the encoding of the satisfied clauses: the objects d are sent out to the skin at the same time the checking is made and they come back to the internal membranes to start the division of these membranes.

(e) $\{[r_{i,k} \to r_{i,k+1}]_2^0 \mid 1 \le i \le m,\ 1 \le k \le 2n-1\}$.
The use of objects r in the rules (i), (j) and (k) makes it necessary to perform a rotation of these objects. This is the mission of the rules of (e).

(f) $\{[d_k \to d_{k+1}]_1^0 \mid n \le k \le 3n-3\}$; $[d_{3n-2} \to d_{3n-1}e]_1^0$.
Through the counter-objects d, these rules *control* the rotation of the elements $r_{i,k}$ in the internal membranes.

(g) $e[\,]_2^0 \to [c_1]_2^+$; $[d_{3n-1} \to d_{3n}]_1^0$.
The application of the rules of (g) will show that the system is ready to check which clauses are true by the assignment encoded by an internal membrane.

(h) $\{[d_k \to d_{k+1}]_1^0 \mid 3n \leq k \leq 3n+2m+2\}$.
These rules supply counters in the skin through objects d in such a way that when objects d_{3n+2m} appear, then this indicates the end of the checking of the clauses. The objects d_k, with $3n+2m+1 \leq k \leq 3n+2m+3$, will control the final stage of the computation.

(i) $[r_{1,2n}]_2^+ \to [\ ]_2^- r_{1,2n}$.
The applicability of the rule (i) encodes the fact that an internal membrane makes true the clause *represented* by the object $r_{1,2n}$, through a change in the sign of its polarization.

(j) $\{[r_{i,2n} \to r_{i-1,2n}]_2^- \mid 1 \leq i \leq m\}$.
Because of the previous rule, we must relabel the values of r representing the different internal membranes. This is made by means of the rule (j).

(k) $r_{1,2n}[\ ]_2^- \to [r_{0,2n}]_2^+$.
By using this rule, the task of making explicit the assignments that make true the clause encoded in that moment of the execution by the object $r_{1,2n}$ is ended.

(l) $\{[c_k \to c_{k+1}]_2^- \mid 1 \leq k \leq m\}$.
The presence of objects c_k (with $2 \leq k \leq m+1$) in the internal membranes shows that the assignments making true every clause are being determined.

(m) $[c_{m+1}]_2^+ \to [\ ]_2^+ c_{m+1}$.
This rule sends to the skin the objects c_{m+1} appearing in the internal membranes.

(n) $[c_{m+1} \to c_{m+2}t]_1^0$.
By using this rule, the objects c_{m+1} in the skin evolve to objects c_{m+2} and t. The objects t in the skin are produced simultaneously with the appearance of the objects $d_{3n+2m+2}$ in the skin, and will show that there exists some assignment making true the formula.

(o) $[t\ ]_1^0 \to [\ ]_1^+ t$.
This rule sends out of the system an object t changing the polarization of the skin to positive (then no object t remaining in the skin is able to evolve). Hence, an object c_{m+2} can exit the skin producing an object Yes. This object is then sent out

to the environment through the rule (p), telling us that the formula is satisfiable and the computation stops.

(p) $[c_{m+2}]_1^+ \rightarrow [\ ]_1^- Yes$.
This rule allows us to change the polarization in the skin to negative in order to ensure that the objects c_{m+2} remaining in it (if any) are not able to continue evolving.

(q) $[d_{3n+2m+3}]_1^0 \rightarrow [\ ]_1^+ No$.
By means of this rule, the object $d_{3n+2m+3}$ only evolves when the skin has a neutral charge (this is the case when the formula is not satisfiable). Then the system will evolve sending to the environment an object No and changing the polarization of the skin to positive, in order to ensure that the objects $d_{3n+2m+3}$ remaining in the skin do not evolve.

Since $\mathbf{PMC}_{\mathcal{R}}$ is closed under complement and polynomial-time reductions, for any class $\mathcal{R}$ of recognizer P systems from Proposition 2.5, the following result is obtained.

Proposition 2.6. $\mathbf{NP} \cup \mathbf{co\text{-}NP} \subseteq \mathbf{PMC}_{\mathcal{AM}(-n)}$.

In the framework of $\mathcal{AM}(+n)$, it has been showed [3] that `QBF-SAT` can be solved in linear time by a family of recognizer P systems with active membranes (without using dissolution rules) and using division rules for elementary and non-elementary membranes. Moreover, in [36], a (deterministic and efficient) algorithm simulating a single computation of any confluent recognizer P system with active membranes has been described. Such P systems can be simulated by a deterministic Turing machine working with exponential space, and spending a time of the order $O(2^{p(n)})$, for some polynomial $p(n)$.

Proposition 2.7. $\mathbf{PSPACE} \subseteq \mathbf{PMC}_{\mathcal{AM}(+n)} \subseteq \mathbf{EXP}$.

Nevertheless, in [39], a characterization of the power efficiency of confluent P systems with active membranes has been given. Specifically, families of these kind of membrane systems solve in polynomial time exactly the class of problems **PSPACE**.

Proposition 2.8. **PSPACE** = $\mathbf{PMC}_{\mathcal{AM}(+n)}$.

Previous results show that the usual framework of P systems with active membranes for solving decision problems is too powerful from the computational complexity point of view. Therefore, it would be interesting to investigate weaker models of P systems with active membranes able to characterize classical complexity classes below **NP** and providing borderlines between efficiency and non-efficiency.

2.5.3. *Polarizationless P systems with active membranes*

In this kind of P systems with active membranes, the membranes of the system have no electrical charges. We denote by $\mathcal{AM}^0(\alpha, \beta)$ the class of all recognizer polarizationless P systems with active membranes such that: (a) if $\alpha = +d$ (respectively, $\alpha = -d$), then dissolution rules are permitted (respectively, forbidden); and (b) if $\beta = +n$ (respectively, $-n$), then division rules for elementary and non-elementary membranes (respectively, only for elementary membranes) are permitted.

At the beginning of 2005, Păun (problem **F** from [23]) wrote:

> *My favorite question (related to complexity aspects in P systems with active membranes and with electrical charges) is that about the number of polarizations. Can the polarizations be completely avoided? The feeling is that this is not possible — and such a result would be rather sound: passing from no polarization to two polarizations amounts to passing from non–efficiency to efficiency.*

This so-called Păun's conjecture can be formally formulated in terms of membrane computing complexity classes as follows: $\mathbf{P} = \mathbf{PMC}_{\mathcal{AM}^0(+d,-n)}$. Currently, this is one of the most relevant open question in Membrane Computing. Nevertheless, some "partial answers" have been found.

Let Π be a recognizer polarizationless P system with active membranes which do not make use of dissolution rules whose set of rules is R. Then, the dependency graph associated with Π is the directed graph $G_\Pi = (V_\Pi, E_\Pi)$ defined as follows:

$$
\begin{aligned}
V_\Pi &= VL_\Pi \cup VR_\Pi \\
VL_\Pi &= \{(a,h) \in \Gamma \times H \mid \exists u \in \Gamma^* \ ([a \to u]_h \in R) \\
&\quad \vee \exists b \in \Gamma \ ([a]_h \to [\ ]_h b \in R) \\
&\quad \vee \exists b \in \Gamma \ \exists h' = ch(h) \ (a[\ ]_{h'} \to [b]_{h'} \in R) \\
&\quad \vee \exists b, c \in \Gamma \ ([a]_h \to [b]_h [c]_h \in R)\} \\
VR_\Pi &= \{(b,h) \in \Gamma \times H \mid \exists a \in \Gamma \ \exists u \in \Gamma^* \ ([a \to u]_h \in R \\
&\quad \wedge b \ \in \mathrm{alph}(u)) \\
&\quad \vee \exists a \in \Gamma \ \exists h' = ch(h) \ ([a]_{h'} \to [\ ]_{h'} b \in R) \\
&\quad \vee \exists a \in \Gamma \ (a[\ ]_h \to [b]_h \in R) \\
&\quad \vee \exists a, c \in \Gamma \ ([a]_h \to [b]_h [c]_h \in R)\} \\
E_\Pi &= \{((a,h),(b,h')) \mid \exists u \in \Gamma^* \ ([a \to u]_h \\
&\quad \in R \wedge b \in \mathrm{alph}(u) \wedge h = h') \\
&\quad \vee ([a]_h \to [\ ]_h b \in R \ \wedge \ h' = f(h)) \\
&\quad \vee (a[\ ]_{h'} \to [b]_{h'} \in R \ \wedge h \ = f(h')) \\
&\quad \vee \exists c \in \Gamma \ ([a]_h \to [b]_h [c]_h \in R \wedge h = h')\}
\end{aligned}
$$

The dependency graph associated with Π verifies the following property: *every accepting computation of* Π *is characterized by the existence of a path in the graph between two specific nodes.* Based on this property and by using the tractability of the *reachability* problem, the following result has been established [10].

Theorem 2.3. $\mathbf{P} = \mathbf{PMC}_{\mathcal{AM}^0(-d,-n)} = \mathbf{PMC}_{\mathcal{AM}^0(-d,+n)}$.

From this theorem, we deduce that polarizationless P systems with active membranes which do not make use of dissolution rules cannot solve **NP**-complete problems in polynomial time (unless $\mathbf{P} = \mathbf{NP}$). This result can be considered as a *partial affirmative* answer to the Păun's conjecture (if dissolution rules are forbidden, then the answer is `yes`).

Let us now consider polarizationless P systems with active membranes making use of dissolution rules. Will it be possible to solve **NP**-complete problems in that framework?

Several authors [1, 10] gave a positive answer when division for non-elementary membranes is allowed. The mentioned papers provide solutions in a linear time to `SAT` and `Subset Sum`, respectively. Thus, we have the following result.

Theorem 2.4. $\mathbf{NP} \cup \mathbf{co\text{-}NP} \subseteq \mathbf{PMC}_{\mathcal{AM}^0(+d,+n)}$.

As a consequence of this theorem, a *partial negative* answer to Păun's conjecture is given: assuming that $\mathbf{P} \neq \mathbf{NP}$ and making use of dissolution rules and division rules for elementary and non-elementary membranes, computationally hard problems can be efficiently solved avoiding polarizations (if division rules for non-elementary membranes are permitted, then the answer is `no`).

Nevertheless, the efficient solvability of **NP**-complete problems by polarizationless P systems with active membranes making use of dissolution rules and division *only* for elementary membranes currently is an open question.

Remark 2.2. In the framework of polarizationless P systems with active membranes, the use or not of the dissolution rules provides a borderline for the tractability of decision problems, assuming that $\mathbf{P} \neq \mathbf{NP}$, that is, by using dissolution rules, we can solve **NP**-complete problems in polynomial time, but without dissolution rules only problems in **P** can be solved in an efficient way.

2.5.4. *Cell-like P systems with symport/antiport rules*

Networks of membranes which compute by communication only, using symport/antiport rules were considered in [25]. By means of this kind of rules, a change of the places of objects with respect to the membranes of a system takes place along computations but not a change/evolution of the objects themselves. Such rules are used both for communication with the environment and for direct communication between different membranes. It is worth noting that, in these systems, the environment plays an active role because we cannot only send objects outside the system, but we can also bring in objects from the environment.

Basic cell-like P systems with symport/antiport rules were introduced in [26] aiming to abstract the biological phenomenon of trans-membrane transport of couples of chemical substances, in the same or in opposite directions.

Definition 2.7. A P system with symport/antiport rules of degree $q \geq 1$ is a tuple $\Pi = (\Gamma, \mathcal{E}, \mu, \mathcal{M}_1, \ldots, \mathcal{M}_q, \mathcal{R}_1, \ldots, \mathcal{R}_q, i_{out})$, where:

- Γ is a finite *alphabet* and $\mathcal{E} \subsetneq \Gamma$,
- μ is a rooted tree whose nodes are injectively labelled with $1, \ldots, q$,
- $\mathcal{M}_1, \ldots, \mathcal{M}_q$ are multisets over Γ,
- $\mathcal{R}_1, \ldots, \mathcal{R}_q$ are finite set of communication rules of the following forms:
  - ⋆ *Symport rules*: (u, out) or (u, in), where u is a finite multiset over Γ such that $|u| > 0$,
  - ⋆ *Antiport rules*: $(u, out; v, in)$, where u, v are finite multisets over Γ such that $|u| > 0$ and $|v| > 0$,
- $i_{out} \in \{0, 1, \ldots, q\}$.

A P system with symport/antiport rules of degree q, can be viewed as a set of q membranes, labelled by $1, \ldots, q$, arranged in a hierarchical (membrane) structure μ given by a rooted tree whose root is called the *skin membrane*, labelled by 1, such that: (a) $\mathcal{M}_i$, $1 \leq i \leq q$, represent the finite multisets of *objects* (symbols of Γ) initially placed in the q membranes of the system; (b) $\mathcal{E}$ is the set of objects initially located in the environment of the system, all of them available in an arbitrary number of copies; (c) $\mathcal{R}_i$, $1 \leq i \leq q$, are finite sets of rules over Γ associated with the membranes labelled by i; (d) i_{out} is a label that represents the output *region* of the system. We use the term *region* i to refer to membrane i, in case $1 \leq i \leq q$, and to refer to the environment, in case $i = 0$. If the alphabet of the environment is an empty set, then we say that the P system is *without environment*. This term means that there is not any object initially located in the environment of the system available in an arbitrary number of copies, that is, in P systems without an environment, there is an environment (labelled by 0 as usual) but in any moment each object in it has a *finite multiplicity*.

A *configuration* at an instant t of a P system with symport/antiport rules is described by the membrane structure at instant t, all multisets of objects over Γ associated with all the membranes present in the system, and the multiset of objects over Γ–$\mathcal{E}$ associated with the environment at that moment. Recall that there are infinite copies of objects from $\mathcal{E}$ in the environment, and hence this set is not properly changed along the computation. The *initial configuration* of the system is $(\mu, \mathcal{M}_1, \ldots, \mathcal{M}_q; \emptyset)$.

A symport rule $(u, out) \in \mathcal{R}_i$ is *applicable* to a configuration $\mathcal{C}_t$ at an instant t if membrane i is in $\mathcal{C}_t$ and multiset u is contained in such a membrane. When applying a rule $(u, out) \in \mathcal{R}_i$, the objects specified by u are sent out of membrane i into the region immediately outside (the parent $p(i)$ of i), this can be the environment in the case of the skin membrane. A symport rule $(u, in) \in \mathcal{R}_i$ is *applicable* to a configuration $\mathcal{C}_t$ at an instant t if membrane i is in $\mathcal{C}_t$ and multiset u is contained in the parent of i. When applying a rule $(u, in) \in \mathcal{R}_i$, the multiset of objects u goes out from the parent membrane of i and enters into the region defined by the membrane i.

An antiport rule $(u, out; v, in) \in \mathcal{R}_i$ is *applicable* to a configuration $\mathcal{C}_t$ at an instant t if membrane i is in $\mathcal{C}_t$ and multiset u is contained in such membrane, and multiset v is contained in the parent of i. When applying a rule $(u, out; v, in) \in \mathcal{R}_i$, the objects specified by u are sent out of membrane i into the parent of i and, at the same time, bringing the objects specified by v into membrane i.

The rules of a P system with symport/antiport rules are applied in a non-deterministic maximally parallel manner: at each step, we apply a multiset of rules which is maximal, no further applicable rule can be added.

Let us fix a P system with symport/antiport rules Π. We say that configuration $\mathcal{C}^1$ yields configuration $\mathcal{C}^2$ in one *transition step*, denoted by $\mathcal{C}^1 \Rightarrow_{\Pi} \mathcal{C}^2$, if we can pass from $\mathcal{C}^1$ to $\mathcal{C}^2$ by applying the rules from $\mathcal{R}_1 \cup \cdots \cup \mathcal{R}_q$ following the previous remarks. A *computation* of Π is a (finite or infinite) sequence of configurations such that: (a) the first term of the sequence is the initial configuration of the system; (b) each non-initial configuration of the sequence is obtained from the previous configuration by applying rules of the system in a maximally parallel manner with the restrictions previously mentioned; and (c) if the sequence is finite (called *halting computation*), then the last term of the sequence is a *halting configuration* (a configuration where no rule of the system is applicable to it).

All computations start from an initial configuration and proceed as stated above; only halting computations give a result, which is encoded by the objects present in the output region i_{out} in the halting configuration. If $\mathcal{C} = \{\mathcal{C}_t\}_{t \le r}$ of Π ($r \in \mathbb{N}$) is a halting computation, then the *length of* $\mathcal{C}$, denoted by $|\mathcal{C}|$, is r, that is, $|\mathcal{C}|$ is the number of non-initial configurations which appear in the finite sequence $\mathcal{C}$. We denote by $\mathcal{C}_t(i), 1 \le i \le q$, the multiset of objects over Γ contained in the membrane labelled by i at configuration $\mathcal{C}_t$. We also denote by $\mathcal{C}_t(0)$ the multiset of objects over $\Gamma \setminus \mathcal{E}$ contained in the environment at configuration $\mathcal{C}_t$.

We denote by $\mathcal{CC}$ the class of all recognizer basic cell-like P systems with symport/antiport rules. What about the efficiency of this kind of P systems? It has been shown that recognizer tissue P systems with symport/antiport rules can be efficiently simulated by basic recognizer P systems (see [5] for details). This proof can be easily adapted to the corresponding cell-like approach. Then, the following proposition holds.

Proposition 2.9. $\mathbf{P} = \mathbf{PMC}_{\mathcal{CC}}$.

In P systems with symport/antiport rules, two mechanisms inspired in cell division and membrane fission processes are introduced by means of two new kind of rules called membrane division rules and membrane separation rules.

A *division rule* is of the form $[a]_i \to [b]_i[c]_i$ and it is *applicable* to a configuration $\mathcal{C}_t$ at an instant t if the following conditions hold: (a) membrane i is in $\mathcal{C}_t$; (b) object a is contained in such a membrane; and (c) membrane i is neither the skin membrane nor the output membrane (if $i_{out} \in \{1, \dots, q\}$). When applying such a rule to configuration $\mathcal{C}_t$, under the influence of object a, the membrane with label i is divided into two membranes with the same label; in the first copy, object a is replaced by object b, in the second one, object a is replaced by object c; all the other objects residing in membrane i are *replicated* and copies of them are placed in the two new membranes.

A *separation rule* is of the form $[a]_i \to [\Gamma_0]_i[\Gamma_1]_i$, where $\{\Gamma_0, \Gamma_1\}$ is a partition of Γ and each separation rule of the system has associated the same partition. Separation rule $[a]_i \to [\Gamma_0]_i[\Gamma_1]_i$ is *applicable* to a configuration $\mathcal{C}_t$ at an instant t if the following conditions hold: (a) membrane i is in $\mathcal{C}_t$; (b) object a is contained in such membrane; and (c) membrane i is neither the skin membrane nor the output membrane (if $i_{out} \in \{1, \dots, q\}$). When applying a separation rule $[a]_i \to [\Gamma_0]_i[\Gamma_1]_i \in \mathcal{R}_i$, in reaction with an object a, the membrane i is separated into two membranes with the same label; at the same time, object a is consumed and all the other objects residing in membrane i are *distributed* as follows: the objects from Γ_0 are placed in the first membrane, those from Γ_1 are placed in the second membrane.

With respect to the semantics of these variants, the rules of such P systems are applied in a non-deterministic maximally parallel manner with the following important remark: when a membrane i is divided (respectively, separated), the division rule (respectively, separation rule) is the only one from $\mathcal{R}_i$ which is applied for that membrane at that step (however, some rules can be applied in a daughter membrane). The new membranes resulting from division (respectively, separation) could participate in the interaction with other membranes or the environment by means of communication rules at the next step — provided that they are not divided (respectively, separated) once again. The label of a membrane precisely identifies the rules which can be applied to it.

The concept of recognizer membrane system is extended to P systems with symport/antiport rules and membrane division or membrane separation in a natural way. We denote by $\mathcal{CDC}(k)$ (respectively, $\mathcal{CSC}(k)$) the class of recognizer P systems with symport/antiport rules and membrane division (respectively, membrane separation) such that the communication rules of the system have length at most k. In the case of P systems with symport/antiport rules and without environment, the following notations will be used: $\widehat{\mathcal{CDC}}(k)$, $\widehat{\mathcal{CSC}}(k)$. If there is no restriction about the length of communication rules, then the term "(k)" will be omitted.

By using the dependency graph technique, the limitations of efficient computations in the previous P systems have been shown in [18] when the communication rules only involve a single object, that is, their length is equal to 1.

Proposition 2.10. $\mathbf{P} = \mathbf{PMC}_{\mathcal{CDC}(1)} = \mathbf{PMC}_{\mathcal{CSC}(1)}$.

By using the *algorithmic technique*, it has been shown [16] that the class $\mathcal{CSC}(2)$ is *feasible*. However, in the cited paper, a polynomial-time solution to the SAT problem by means of a family of membrane systems from $\mathcal{CSC}(3)$ was given.

Proposition 2.11. $\mathbf{P} = \mathbf{PMC}_{\mathcal{CSC}(2)}$ *and* $\mathbf{NP} \cup \mathbf{co\text{-}NP} \subseteq \mathbf{PMC}_{\mathcal{CSC}(3)}$.

From a computational complexity point of view, what happens when division rules or separation rules are allowed? With respect to division rules, the use of communication rules with length at most two is enough to reach the presumably efficiency. Specifically, in [40], a polynomial-time solution of `HAM-CYCLE` problem by means of a family from $\mathcal{CDC}(2)$ was provided.

Proposition 2.12. $\mathbf{NP} \cup \mathbf{co\text{-}NP} \subseteq \mathbf{PMC}_{\mathcal{CDC}(2)}$.

Let us recall that in the definition of P systems with symport/antiport rules, the objects in the alphabet of the environment appear at the initial configuration of the system in an arbitrarily large number of copies. This property seems an unfair tool when designing efficient solutions to computationally hard problems in the framework of membrane computing by performing a space–time trade-off. In [15], a deterministic algorithm working in polynomial time that receives as input a membrane system Π from $\widehat{\mathcal{CSC}}$ together with an input multiset m of Π in such a manner that algorithm A reproduces the behaviour of a single computation of the system $\Pi + m$. Therefore, the use of algorithmic technique has shown the feasibility of the systems from $\widehat{\mathcal{CSC}}$.

Proposition 2.13. $\mathbf{P} = \mathbf{PMC}_{\widehat{\mathcal{CSC}}}$.

Therefore, in the framework of cell-like P systems with symport/antiport rules and membrane separation, the role of the environment is *relevant* in the following sense: with environment, the systems are presumably efficient (**NP**-complete problems can be solved efficiently) but without environment, they are feasible, that is, only tractable problems can be solved in polynomial time.

What happens if membrane division rules are considered instead of membrane separation rules? Now, the situation is quite different. In [17], each family of systems from $\mathcal{CDC}(k)$ solving a decision problem in polynomial time is simulated in an efficient way by a family of systems from $\widehat{\mathcal{CDC}}(k)$. Thus, we have the following proposition.

Proposition 2.14. $\mathbf{PMC}_{\mathcal{CDC}(k)} = \mathbf{PMC}_{\widehat{\mathcal{CDC}}(k)}$ *for all* $k \geq 1$.

Thus, in the framework of cell-like P systems with symport/antiport rules and membrane division, the role of the environment is *irrelevant* in the following sense: families of recognizer P systems with symport/antiport rules and membrane division whose communication rules are at most $k \geq 1$, both in the case where there is an environment and in the case where there is no environment, solve in polynomial time the same class of decision problems.

2.6. Efficiency of Tissue-Like Membrane Systems

In this section, we study the efficiency of different models of tissue-like membrane systems. These variants are considered by using specific kind of rules. The basic one only considers communication rules but we also analyze tissue P systems where division or separation rules are also permitted.

Tissue P systems with symport/antiport rules were introduced in [8] by abstracting networks of elementary membranes such that some of them are linked by "communication channels".

2.6.1. *Basic tissue P systems*

Definition 2.8. A basic tissue P system with symport/antiport rules of degree $q \geq 1$ is a tuple $\Pi = (\Gamma, \mathcal{E}, \mathcal{M}_1, \ldots, \mathcal{M}_q, \mathcal{R}, i_{out})$, where:

- Γ is a finite *alphabet* and $\mathcal{E} \subsetneq \Gamma$,
- $\mathcal{M}_1, \ldots, \mathcal{M}_q$ are multisets over Γ,
- $\mathcal{R}$ is a finite set of communication rules of the form $(i, u/v, j)$, for $i, j \in \{0, 1, 2, \ldots, q\}, i \neq j$, u, v are finite multisets over Γ such that $|u + v| > 0$,
- $i_{out} \in \{0, 1, 2, \ldots, q\}$.

A *tissue P system with symport/antiport rules* $\Pi = (\Gamma, \mathcal{E}, \mathcal{M}_1, \ldots, \mathcal{M}_q, \mathcal{R}, i_{out})$ of degree $q \geq 1$ can be viewed as a set of q cells, labelled by $1, \ldots, q$, with an environment labelled by 0 such that: (a) $\mathcal{M}_1, \ldots, \mathcal{M}_q$ are strings over Γ representing the finite multisets of objects (elements in Γ) initially placed in the q cells of the system; (b) $\mathcal{E}$ is the set of objects located initially in the environment of the system, all of them appearing in an *arbitrary number of copies*; and (c) $i_{out} \in \{0, 1, 2, \ldots, q\}$ represents a distinguished *region* (a cell in the case $i_{out} \in \{1, 2, \ldots, q\}$ and the environment in the case $i_{out} = 0$) which will encode the output of the system. The *length* of the communication rule $(i, u/v, j)$ is defined as $|u| + |v|$, that is, the total number of objects which appear in the rule.

It is interesting to highlight some differences between the cell-like approach and tissue-like approach. First, in the cell-like approach, only the skin membrane can communicate with the environment but in tissue-like systems each cell can communicate with the environment. Second, in the cell-like approach, two membranes can only communicate if one is the parent of the other but in tissue-like systems any two cells can communicate with each other.

A *configuration* at an instant t of a tissue P system is described by all multisets of objects over Γ associated with all the cells present in the system, and the multiset of objects over Γ–$\mathcal{E}$ associated with the environment at that moment. Bearing in mind that the objects from $\mathcal{E}$ have infinite copies in the environment, they are not properly changed along the computation. The *initial configuration* is $(\mathcal{M}_1, \ldots, \mathcal{M}_q; \emptyset)$. A configuration is a *halting configuration* if no rule of the system is applicable to it.

A communication rule $(i, u/v, j)$ is *applicable* to a configuration $\mathcal{C}_t$ at an instant t if multiset u is contained in region i of $\mathcal{C}_t$ and multiset v is contained in region j of $\mathcal{C}_t$. When applying a rule $(i, u/v, j)$ to configuration $\mathcal{C}_t$, the objects of the multiset represented by u are sent from region i to region j and, simultaneously, the objects of multiset v are sent from region j to region i.

A communication rule $(i, u/v, j)$ is called a *symport rule* if $u = \lambda$ or $v = \lambda$. A symport rule $(i, u/\lambda, j)$, with $i \neq 0, j \neq 0$, provides a virtual arc from cell i to cell j. A communication rule $(i, u/v, j)$ is called an *antiport rule* if $u \neq \lambda$ and $v \neq \lambda$. An antiport rule $(i, u/v, j)$, with $i \neq 0, j \neq 0$, provides two arcs: one from cell i to cell j and another one from cell j to cell i. Thus, every tissue P system has an underlying directed graph whose nodes are the cells of the system and the arcs are obtained from communication rules. In this context, the environment can be considered as a virtual node of the graph such that their connections are defined by communication rules of the form $(i, u/v, j)$ with $i = 0$ or $j = 0$.

The rules of a system like the one above are used in a nondeterministic maximally parallel manner as it is customary in Membrane Computing. At each step, all cells which can evolve must evolve in a maximally parallel way (at each step, we apply a multiset of rules which is maximal, no further applicable rule can be added). Then, the concepts of *transition step* and *computation* are defined in a natural way.

We denote by $\mathcal{TC}$ the class of all recognizer basic tissue P systems. In [5], families of recognizer basic tissue P systems which solve a decision problem can be efficiently simulated by a family of recognizer basic P systems solving the same problem. Thus, bearing in mind that the class $\mathcal{T}$ is feasible, we deduce that the class $\mathcal{TC}$ is also feasible.

Theorem 2.5. $\mathbf{P} = \mathbf{PMC}_{\mathcal{TC}}$.

2.6.2. *Tissue P systems with cell division or cell separation*

Similar to the case of cell-like approach, the mechanisms inspired in cell division and membrane fission processes are introduced in tissue-like approach by means of cell division rules and cell separation rules.

A *division rule* is of the form $[a]_i \to [b]_i[c]_i$ and it is *applicable* to a configuration $\mathcal{C}_t$ at an instant t if the following conditions hold: (a) cell i is in $\mathcal{C}_t$; (b) object a is contained in such a cell; and (c) cell i is not the output membrane (if $i_{out} \in \{1, \ldots, q\}$). When applying such a rule to configuration $\mathcal{C}_t$, under the influence of object a, the cell with label i is divided into two cells with the same label; in the first copy, object a is replaced by object b, in the second one, object a is replaced by object c; all the other objects residing in cell i are *replicated* and copies of them are placed in the two new cells.

A *separation rule* is of the form $[a]_i \to [\Gamma_0]_i[\Gamma_1]_i$, where $\{\Gamma_0, \Gamma_1\}$ is a partition of Γ and each separation rule of the system has associated the same partition. Separation rule $[a]_i \to [\Gamma_0]_i[\Gamma_1]_i$ is *applicable* to a configuration $\mathcal{C}_t$ at an instant t if the following conditions hold: (a) cell i is in $\mathcal{C}_t$; (b) object a is contained in such a membrane; and (c) membrane i is not the output cell (if $i_{out} \in \{1, \ldots, q\}$). When applying a separation rule $[a]_i \to [\Gamma_0]_i[\Gamma_1]_i \in \mathcal{R}_i$, in reaction with an object a, the cell i is separated into two cells with the same label;

at the same time, object a is consumed and all the other objects residing in membrane i are *distributed* as follows: the objects from Γ_0 are placed in the first cell, those from Γ_1 are placed in the second cell.

With respect to the semantics of these variants, the rules of such P systems are applied in a non-deterministic maximally parallel manner with the following important remark: when a cell i is divided (respectively, separated), the division rule (respectively, separation rule) is the only one from $\mathcal{R}_i$ which is applied for that cell at that step (however, some rules can be applied in a daughter membrane). The new cells resulting from division (respectively, separation) could participate in the interaction with other cells or the environment by means of communication rules at the next step — providing that they are not divided (respectively, separated) once again. The label of a cell precisely identifies the rules which can be applied to it.

The concept of recognizer membrane system is extended to tissue P systems with symport/antiport rules and cell division or cell separation in a natural way. We denote by $\mathcal{TDC}(k)$ (respectively, $\mathcal{TSC}(k)$) the class of recognizer tissue P systems with symport/antiport rules and cell division (respectively, cell separation) such that the communication rules of the system have length at most k.

We denote by $\mathcal{TSS}$ (respectively, $\mathcal{TSA}$) the class of recognizer tissue P systems with cell separation and symport rules (respectively, antiport rules). For each natural number $k \geq 1$, we denote by $\mathcal{TSS}(k)$ (respectively, $\mathcal{TSA}(k)$), the class of recognizer tissue P systems with cell separation and symport rules (respectively, antiport rules) of length at most k.

By using the concept of dependency graph associated with tissue P systems with cell division and communication rules with length at most 1, it has been proved that this kind of membrane systems can only efficiently solve tractable problems (see [8] for details).

Theorem 2.6. $\mathbf{P} = \mathbf{PMC}_{\mathcal{TDC}(1)}$.

In [38], a polynomial time solution of the `HAM-CYCLE` problem was given by using a family of recognizer tissue P systems with cell division and communication rules of length at most 2.

Proposition 2.15. $\mathbf{NP} \cup \textbf{co-NP} \subseteq \mathbf{PMC}_{\mathcal{TDC}(2)}$.

In [6], a polynomial time solution of the `Vertex Cover` problem by a family of tissue P systems from $\mathcal{TDC}(3)$ is shown. Besides, the symport rules that are used in that solution have length at most 2. Then, we can rearrange the rules by adding a new object # to the alphabet $\mathcal{E}$ of the environment, and by replacing rules $(i, u/\lambda, 0)$ by rules $(i, u/\#, 0)$. Bearing in mind that `Vertex Cover` is an **NP**-complete problem we have the following result.

Proposition 2.16. $\mathbf{NP} \cup \textbf{co-NP} \subseteq \mathbf{PMC}_{\mathcal{TDA}(3)}$.

Remark 2.3. From Theorem 2.6 and Proposition 2.15, we deduce that in the framework of recognizer tissue P systems with cell division the length of the communication rules provides a borderline of the tractability of decision problems. Specifically, passing from length 1 to length 2 amounts to passing from non-efficiency to efficiency, assuming that $\mathbf{P} \neq \mathbf{NP}$. Moreover, another frontier is obtained when the use of cell division rules in basic tissue P systems is allowed.

By using the concept of dependency graph associated with tissue P systems with cell separation and communication rules with length at most 1, it has been proved that this kind of membrane systems can only efficiently solve tractable problems (see [19] for details). This result has been extended to tissue P systems with cell separation and communication rules with length at most two (see [20] for details).

Theorem 2.7. $\mathbf{P} = \mathbf{PMC}_{\mathcal{TSC}(2)}$.

In [35], a polynomial time solution of the `SAT` problem was given by using a family of recognizer tissue P systems with cell division and communication rules of length at most 3.

Proposition 2.17. $\mathbf{NP} \cup \textbf{co-NP} \subseteq \mathbf{PMC}_{\mathcal{TSC}(3)}$.

Remark 2.4. From Theorem 2.7 and Proposition 2.17, we deduce that in the framework of recognizer tissue P systems with cell division the length of the communication rules provides a borderline of the tractability of decision problems. Specifically, passing from length

1 to length 3 amounts to passing from non-efficiency to efficiency, assuming that $\mathbf{P} \neq \mathbf{NP}$.

In the solution provided in [35], the symport rules that appear are of the type $(i, u/\lambda, 0)$ with $|u| \leq 2$. Then we can rearrange the rules adding a new object # to the alphabet $\mathcal{E}$ of the environment in such a way that symport rules $(i, u/\lambda, 0)$ can be replaced by equivalent antiport rules $(i, u/\#, 0)$. This proves that $\texttt{SAT} \in \mathbf{PMC}_{\mathcal{TSA}(3)}$.

Proposition 2.18. $\mathbf{NP} \cup \textbf{co-NP} \subseteq \mathbf{PMC}_{\mathcal{TSA}(3)}$.

In [13], the role of the direction in communication rules of tissue-like P systems with cell separation is analyzed from a computational complexity point of view, in the case when their communication rules are all of the same type: either symport rules or antiport rules. Specifically, the following result has been shown.

Proposition 2.19. $\mathbf{P} = \mathbf{PMC}_{\mathcal{TSS}(k)}$ *for each* $k \geq 1$.

2.6.3. *Tissue P systems without environment*

Next, we analyze the role of the environment in tissue P systems with symport/antiport rules and with cell division or with cell separation from the complexity point of view.

Let us recall that a *tissue P system without environment* is a tissue P system such that the alphabet of the environment is an empty set. For each natural number $k \geq 1$, we denote by $\widehat{\mathcal{TDC}}(k)$ (respectively, $\widehat{\mathcal{TSC}}(k)$) the class of recognizer tissue P systems with cell division (respectively, tissue P systems with cell separation) and with communication rules of length at most k.

In [33], it is showed that every family $\mathbf{\Pi}$ of tissue P systems with cell division solving a decision problem can be simulated by a family $\mathbf{\Pi}'$ of tissue P system with cell division and without environment in an efficient way. Moreover, the upper bound of the length of communication rules of the systems from $\mathbf{\Pi}$ is equal to the maximum length of communication rules of the system from $\mathbf{\Pi}'$. That is, we have the following result.

Theorem 2.8. *For each* $k \geq 1$, *we have* $\mathbf{PMC}_{\mathcal{TDC}(k)} = \mathbf{PMC}_{\widehat{\mathcal{TDC}}(k)}$.

From this theorem, we deduce that the environment is not relevant for the presumable efficiency of tissue P systems with cell division.

Concerning the role of the environment in tissue P systems with cell separation, in [14], by using the algorithmic technique it has been shown that only problems in class **P** can be solved efficiently by using tissue P systems with communication rules, separation rules and without environment. That is, we have the following result.

Theorem 2.9. *For each* $k \geq 1$, *we have* $\mathbf{P} = \mathbf{PMC}_{\widehat{\mathcal{TSC}}(k)}$.

From this theorem, we deduce that the environment is relevant for the efficiency of tissue P system with cell separation.

Remark 2.5. From Theorem 2.9 and Proposition 2.15, we deduce that in the framework of recognizer tissue P systems with cell separation, passing from to-have environment or not amounts to passing from efficiency to non-efficiency, assuming that $\mathbf{P} \neq \mathbf{NP}$. Moreover, another frontier is obtained when we allow to use separation cell in basic tissue P systems.

Remark 2.6. Bearing in mind that **NP**-complete problems can be efficiently solved by using tissue P systems with cell division and without environment, we deduce that in the framework of tissue P systems without environment, the kind of rules (separation versus division) provides a new frontier of the tractability of decision problems, assuming that $\mathbf{P} \neq \mathbf{NP}$.

2.7. Frontiers of the Efficiency in Membrane Computing

Next, based on the results from previous sections, different frontiers of the efficiency in terms of syntactical ingredients of recognizer membrane systems are presented. It is worth noting that each borderline provides a new way to tackle the **P** versus **NP** problem.

Cell-like membrane systems

Feasible	Presumably Efficient	Ingredient
$\mathcal{NAM}$	$\mathcal{AM}$	(*division rules*)
$\mathcal{AM}^0(-d,-n)$	$\mathcal{AM}(-d,-n)$	(*polarizations*)
$\mathcal{AM}^0(-d,+n)$	$\mathcal{AM}^0(+d,+n)$	(*dissolution rules*)
$\mathcal{CC}$	$\mathcal{CDC}$	(*division rules*)
$\mathcal{CC}$	$\mathcal{CSC}$	(*separation rules*)
$\mathcal{CDC}(1)$	$\mathcal{CDC}(2)$	(*length*)
$\mathcal{CSC}(2)$	$\mathcal{CSC}(3)$	(*length*)
$\mathcal{CSC}(2)$	$\mathcal{CDC}(2)$	(*kind of rules*)
$\widehat{\mathcal{CSC}}$	$\mathcal{CSC}$	(*environment*)
$\widehat{\mathcal{CSC}}(3)$	$\mathcal{CSC}(3)$	(*environment*)

Tissue-like membrane systems

Feasible	Presumably Efficient	Ingredient
$\mathcal{TC}$	$\mathcal{TDC}$	(*division rules*)
$\mathcal{TC}$	$\mathcal{TSC}$	(*separation rules*)
$\mathcal{TDC}(1)$	$\mathcal{TDC}(2)$	(*length*)
$\mathcal{TSC}(2)$	$\mathcal{TSC}(3)$	(*length*)
$\mathcal{TSC}(2)$	$\mathcal{TDC}(2)$	(*kind of rules*)
$\widehat{\mathcal{TSC}}$	$\mathcal{TSC}$	(*environment*)
$\widehat{\mathcal{TSC}}$	$\widehat{\mathcal{TDC}}$	(*kind of rules*)
$\widehat{\mathcal{TSC}}(3)$	$\mathcal{TSC}(3)$	(*environment*)
$\mathcal{TSS}$	$\mathcal{TSC}$	(*direction*)
$\mathcal{TSS}$	$\mathcal{TSA}$	(*direction*)
$\mathcal{TSS}(3)$	$\mathcal{TSA}(3)$	(*direction*)
$\mathcal{TSA}(2)$	$\mathcal{TSA}(3)$	(*length*)

Bibliography

[1] Alhazov, A., Pan, L. and Păun, Gh. Trading polarizations for labels in P systems with active membranes. *Acta Inform.* **41**(2–3), 111–144, 2004.

[2] Alhazov, A., Martín-Vide, C. and Pan, L. Solving graph problems by P systems with restricted elementary active membranes. *Lect. Notes Comput. Sci.* **2950**, 1–22, 2004.

[3] Alhazov, A., Martín-Vide, C. and Pan, L. Solving a PSPACE — complete problem by recognizing P systems with restricted active membranes. *Fundam. Inform.* **58**, 67–77, 2003.

[4] Cormen, T. H., Leiserson, C. E. and Rivest, R. L. *An Introduction to Algorithms.* MIT Press, Cambridge, MA, 1994.

[5] Díaz-Pernil, D., Gutiérrez-Naranjo, M. A., Pérez-Jiménez, M. J. and Romero-Jiménez, A. Efficient simulation of tissue-like P systems by transition cell-like P systems. *Nat. Comput.* **8**(4), 797–806, 2009.

[6] Díaz-Pernil, D., Pérez-Jiménez, M. J., Riscos-Núñez, A. and Romero-Jiménez, A. Computational efficiency of cellular division in tissue-like membrane systems. *Romanian J. Inf. Sci. Technol.* **11**(3), 229–241, 2008.

[7] Goldreich, O. *Computational Complexity: A Conceptual Perspective*, Cambridge University Press, New York, 2008.

[8] Gutiérrez-Escudero, R., Pérez-Jiménez M. J. and Rius-Font, M. Characterizing tractability by tissue-like P systems. *Lect. Notes Comput. Sci.* **5957**, 289–300, 2010.

[9] Gutiérrez-Naranjo, M. A., Pérez-Jiménez M. J. and Riscos-Núñez, A. A fast P system for finding a balanced 2-partition. *Soft Comput.* **9**(9), 673–678, 2005.

[10] Gutiérrez-Naranjo, M. A., Pérez-Jiménez, M. J., Riscos-Núñez, A. and Romero-Campero, F. J. On the power of dissolution in P systems with active membranes. *Lect. Notes Comput. Sci.* **3850**, 224–240, (2006).

[11] Gutiérrez-Naranjo, M. A., Pérez-Jiménez, M. J. Riscos-Núñez, Romero-Campero A. F. J. and Romero-Jiménez, A. Characterizing tractability by cell-like membrane systems. In K. G. Subramanian, K. Rangarajan and M. Mukund (eds.). *Formal Models, Languages and Applications*, pp. 137–154, World Scientific, Singapore, 2006.

[12] Ladner, R. E. On the structure of polynomial time reducibility. *J. Assoc. Comput. Mach.* **22**(1), 155–171, 1975.

[13] Macías-Ramos, L. F., Martínez-del-Amor, M. A., Pérez-Jiménez, M. J. and Valencia-Cabrera, L. The role of the direction in tissue P systems with cell separation. *J. Autom. Lang. Combin.* **19**(1–4), 185–199, 2014.

[14] Macías-Ramos, L. F., Pérez-Jiménez, M. J., Riscos-Núñez, A. and Rius-Font, M. The efficiency of tissue P systems with cell separation relies on the environment. *Lect. Notes Comput. Sci.* **7762**, 243–256, 2013.

[15] Macías-Ramos, L. F., Pérez-Jiménez, M. J., Riscos-Núñez, A. and Valencia-Cabrera, L. Membrane fission versus cell division: when membrane proliferation is not enough. *Theoret. Comput. Sci.* **608**, 57–65, 2015.

[16] Macías-Ramos, L. F., Song, B., Valencia-Cabrera, L., Pan, L. and Pérez-Jiménez, M. J. Membrane fission: A computational complexity perspective. *Complexity* **21**(6) 321–334, 2016.
[17] Macías-Ramos, L. F., Song, B., Song, T., Pan, L. and Pérez-Jiménez, M. J. On efficiency of P systems with symport/antiport and membrane division, BWMC 2017: 15th Brainstorming Week on Membrane Computing, 129–146, 2017.
[18] Macías-Ramos, L. F., Valencia-Cabrera, L., Song, B., Song, T., Pan, L. and Pérez-Jiménez, M. J. Limits on efficient computation in P systems with symport/antiport rules. BWMC 2017: 15th Brainstorming Week on Membrane Computing, 147–160, 2017.
[19] Pan, L. and Pérez-Jiménez M. J. Computational complexity of tissue-like P systems. *J. Complexity* **26**(3), 296–315, 2010.
[20] Pan, L., Pérez-Jiménez, M. J., Riscos, A. and Rius, M. New frontiers of the efficiency in tissue P systems. In L. Pan, Gh. Paun and T. Song (eds.). *Preproceedings of Asian Conference on Membrane Computing*, pp. 61–73, 2012, Huazhong University of Science and Technology, Wuhan, China, October 5–18.
[21] Păun, Gh. P systems with active membranes: Attacking **NP**-complete problems. *J. Autom. Lang. Combin.* **6**(1), 75–90, 2001.
[22] Păun, Gh. *Membrane Computing: An Introduction.* Springer-Verlag, Berlin, 2002.
[23] Păun, Gh. Further twenty six open problems in membrane computing. In M. A. Gutiérrez *et al.* (eds.). *Third Brainstorming Week on Membrane Computing*, pp. 249–262. Fénix Editora, Sevilla, 2005.
[24] Păun, Gh., Pérez-Jiménez, M. J. and Riscos-Núñez, A. Tissue P system with cell division. *Int. J. Comput. Commun. Control* **3**(3), 295–303, 2008.
[25] Păun, A., Păun, Gh. and Rozenberg, G. Computing by communication in networks of membranes, *Int. J. Found. Comput. Sci.* **13**(6), 779–798, 2002.
[26] Păun, A. and Păun, Gh. The power of communication P systems with symport/antiport, *New Gener. Comput.* **20**(3), 295–305, 2002.
[27] Pérez-Jiménez, M. J., Romero-Jiménez, A. and Sancho-Caparrini, F. Complexity classes in models of cellular computing with membranes. *Nat. Comput.* **2**(3), 265–285, 2003.
[28] Pérez-Jiménez, M. J. and Riscos-Núñez, A. A linear-time solution to the Knapsack problem using P systems with active membranes. *Lect. Notes Comput. Sci.* **2933**, 250–268, 2004.
[29] Pérez-Jiménez M. J. and Riscos-Núñez, A. Solving the subset-sum problem by active membranes. *New Gener. Comput.* **23**(4), 367–384, 2005.
[30] Pérez-Jiménez M. J. and Romero-Campero, F. J. An efficient family of P systems for packing items into bins. *J. Universal Comput. Sci.* **10**(5), 650–670, 2004.
[31] Pérez-Jiménez, M. J. and Romero-Campero, F. J. Attacking the common algorithmic problem by recognizer P systems. *Lect. Notes Comput. Sci.* **3354**, 304–315, 2005.

[32] Pérez-Jiménez, M. J., Romero-Jiménez, A. and Sancho-Caparrini, F. A polynomial complexity class in P systems using membrane division. *J. Autom. Lang. Combin.* **11**(4), 423–434, 2006.

[33] Pérez-Jiménez, M. J., Riscos-Núñez, A., Rius-Font M. and Romero-Campero, F. J. A polynomial alternative to unbounded environment for tissue P systems with cell division. *Int. J. Comput. Math.* **90**(4), 760–775, 2013.

[34] Pérez-Jiménez, M. J. The P versus NP problem from membrane computing view. *Eur. Rev.* **22**(01), 18–33, 2014.

[35] Pérez-Jiménez M. J. and Sosík, P. An optimal frontier of the efficiency of tissue P systems with cell separation. *Fundam. Inform.* **138**(1–2), 45–60, 2015.

[36] Porreca, A. E., Mauri, G. and Zandron, C. Complexity classes for membrane systems. *RAIRO — Theor. Inf. Appl.* **40**(2), 141–162, 2006.

[37] Porreca, A. E. (2008). *Computational Complexity Classes for Membrane Systems.* Master Thesis, Universita' di Milano-Bicocca, Italy.

[38] Porreca, A. E., Murphy, N. and Pérez-Jiménez, M. J. An optimal frontier of the efficiency of tissue P systems with cell division. In M. Garcí-Quismondo, L. F. Macías-Ramos, Gh. Paun, I. Pérez Hurtado and L. Valencia-Cabrera (eds.). *Proceedings of the Tenth Brainstorming Week on Membrane Computing, Volume II*, pp. 141–166, 2012, Seville, Spain, January 30–February 3, Report RGNC 01/2012, Fénix Editora.

[39] Sosík, P. and Rodríguez-Patón, A. Membrane computing and complexity theory: A characterization of PSPACE. *J. Comput. Syst. Sci.* **73**, 137–152, 2007.

[40] Valencia-Cabrera, L. Song, B., Macías-Ramos, L. F., Pan, L., Riscos-Núñez A. and Pérez-Jiménez, M. J. Minimal cooperation in P systems with symport/antiport: A complexity approach. In L. F. Macías, Gh. Păun, A. Riscos and L. Valencia (eds.). *Proceedings of the Thirteenth Brainstorming Week on Membrane Computing*, pp. 301–323, 2015, 2–6 February, Sevilla, Spain.

[41] Zandron, C., Ferretti, C. and Mauri, G. Solving NP-complete problems using P systems with active membranes. In I. Antoniou, C. S. Calude and M. J. Dinneen (eds.). *Unconventional Models of Computation*, pp. 289–301, 2000, Springer, Berlin.

Chapter 3

Integrative Approaches for Predicting microRNA Function and Prioritizing Disease-Related microRNA Using Biological Interaction Networks

Xuan Zhang* and Xiangxiang Zeng*,†,‡

*School of Information Science and Technology
Xiamen University, Xiamen, China
†Department of Artificial Intelligence
Faculty of Computer Science
Polytechnical University of Madrid
Campus de Montegancedo
Boadilla del Monte 28660, Madrid, Spain
‡xzeng@xmu.edu.cn

MicroRNAs (miRNAs) play critical roles in regulating gene expressions at post-transcriptional levels. The prediction of disease-related miRNA is vital to the further investigation of miRNA's involvement in the pathogenesis of a disease. In previous years, biological experimentation was the main method used to identify whether miRNA is associated with a given disease. With increasing biological information and the appearance of new miRNAs every year, experimental identification of disease-related miRNAs poses considerable difficulties (e.g. time consumption and high cost). Because of the experimental methods' limitations in determining the relationship between miRNAs and diseases, computational methods have been proposed. A key to predict potential disease-related miRNA based on networks is the calculation of similarity among diseases and miRNA over the networks. Different strategies lead to different results. In this chapter, we summarize the existing computational approaches and present the confronted difficulties that help understand the research status. We also discuss the principles, efficiency and differences among these methods. The comprehensive comparison and discussion elucidated in this work provide constructive insights into the matter.

3.1. Introduction

MicroRNAs (miRNAs) are a class of small, endogenous, single-stranded, non-coding RNAs (~22 nucleotides) that mainly repress the expression of target mRNAs at the post-transcriptional level by binding to the 3′-UTR of target mRNAs through sequence-specific base pairing, resulting in target mRNAs cleavage or translation inhibition [2, 3, 29]. Increasing evidence suggests that miRNAs may function as positive regulators at the post-transcriptional level, which is critical for development of diseases [21, 42]. The lin-4 and let-7 in *Caenorhabditis elegans* were the first miRNAs to be discovered [37]. Since then, using various experimental methods, numerous miRNAs have been identified. As of June 2014, investigators have discovered and documented 28,645 miRNA entries, which can be run through the miRBase. Increasing evidence suggests that the mutation of miRNA, the dysfunction of miRNA biogenesis and the dysregulation of miRNAs and miRNA target genes may lead to various diseases. As such, miRNA, first discovered in *Caenorhabditis elegans* lin-4, encodes a 22-nt RNA fragment length and regulates the expression of its target genes lin-14 [44] and lin-28 [32] and plays an important regulatory role in the development of nematode larvae [37]. The second-discovered miRNA let-7 regulates its target gene lin-41 [40] and hbl-1 [1].

Herein, identifying the interactions between miRNAs and diseases is a crucial problem. Applying only experimental methods to discover such a correlation poses many bottlenecks such as lengthy experimental periods, high equipment requirements and high cost. With the emergence of numerous miRNAs, various databases have been presented to store meaningful information about these RNA molecules. Based on existing data, as a supplement of experimental methods, computational approaches were proposed to solve the problem. Computational means obtain the potential associations between miRNAs and diseases in a short time and greatly reduce the experimental workload. However, many challenges also confront these novel approaches. First, we have only positive samples without negative samples. Because, through several experimental means, we can prove that a miRNA is related with a disease, but cannot prove that a miRNA is absolutely unrelated with a disease. This results in fewer positive associations and a vast amount of unlabelled samples. Second, when a new miRNA is discovered, there is no information about it so that the computational approaches can hardly predict the relationship of this RNA molecule with any disease.

The rest of this chapter is organized as follows. Section 3.2 summarizes the databases used to predict certain miRNA–disease associations; these databases can be downloaded from the specified links. Section 3.3 reviews previous approaches and divides them into two categories. The first category is based on the similarity network method; the other is based on the machine learning method. Finally, Section 3.4 concludes the chapter and discusses the advantages and disadvantages of previous approaches.

3.2. Database

In this section, we present related databases to help predict certain miRNA–disease associations.

3.2.1. *miRNA–disease network*

HMDD v2.0 and miR2Disease are databases that have long been developed and made publicly available. These information banks can be accessed through a web interface at http://cmbi.bjmu.edu.cn/hmdd and http://www.miR2Disease.org/, respectively. HMDD v2.0 is a database to experimentally support human miRNA and disease associations [26]. MiR2Disease is a manually curated database for miRNA deregulation in human disease [17]. Another special database, miRCancer, focuses on the miRNA–cancer relationship. The miRCancer is a miRNA–cancer association database constructed by text mining the literature [45]. The dbDEMC is a database of differentially expressed miRNAs in human cancers [50].

3.2.2. *Disease phenotype network*

To obtain more comprehensive and complete networks, some researchers use the disease phenotype network, which is based on more information and easily attains good performance. Accordingly, certain scholars, via the gene–phenotype network, obtain a disease similar network. A more recent database for disease and gene association, named DGA, can be downloaded from http://dga.nubic.northwestern.edu [35]. Genes associated with similar diseases show both the high likelihood of physical interactions and the higher expression profiling similarity for their transcripts, supporting the existence of distinct disease-specific functional modules [11]. Theoretically, this guarantees the feasibility of the aforementioned strategy. More generally, by text mining, the MimMiner database is widely used to classify over 5000 human phenotypes contained in the Online Mendelian Inheritance in Man database. Researchers found that the similarity between phenotypes reflects biological modules of interacting functionally related genes [41]. The phenotype association data can be downloaded at http://www.cmbi.ru.nl/MimMiner/. In addition, another database, MeSH, which is available at http://www.ncbi.nlm.nih.gov/, can be translated into disease sematic similarity information through a process to be described later [43].

3.2.3. *miRNA association networks*

The miRNA–miRNA relationship cannot be directly identified. Therefore, establishing a functional association between two different miRNAs, generally via their target–genes network, is useful for attaining good performance. Numerous miRNA–target gene databases are adopted. We roughly divide these databases into two categories arbitrarily designated, “Category A” and “Category B”, according to the source of their relation.

Category A: Database miRNAMap 2.0, a representative class in this category, contains experimentally verified miRNAs and miRNA target genes in human, mouse, rat and other metazoan genomes [43]. The miRNAMap 2.0 is now available at http://miRNAMap.mbc.nctu.edu.tw/. TarBase [38] and miRNA.org [5], which also belong to this category, can be download from http://www.diana.pcbi.upenn.edu/tarbase/ and http://www.miRNA.org/.

Category B: Another database, miRBase Sequences, is the primary online repository for miRNA sequence data and annotation, which is a comprehensive new database of predicted miRNA target genes [12]. MiRBase is available at http://miRNA.sanger.ac.uk/. Similar databases in this category, such as PITA [22], miRGator [33], miRGen [28], PicTar [23], TargetScan [14], DIANA-microT [27], RNAhybrid [24] and RNA22 [30], are often used to obtain the relationship between miRNA and its target genes. All these databases are available online. Based on prior work, the miRNA–miRNA functional similarity scores can be downloaded from http://cmbi.bjmu.edu.cn/misim/ [43]. Rfam [13], an RNA family database, may also be used when miRNA familiar information is considered. Besides, the miRBase [12] also includes integrated interfaces to comprehensive microRNA sequence data and annotation, which can be used to compute the similarity among miRNAs.

3.2.4. *Gene interaction network*

Genes play an extremely important role in biological inheritance. To a certain extent, genes with similar functions are often associated with similar diseases. For this reason, a large amount of gene information is used to predict diseases. Gene–phenotype relationships, which can be obtained from the OMIM database, are widely used [16]. Because OMIM does not routinely collect findings of lower significance or negative findings and because of the increase in the development of genetic association databases, GAD was built [4]. The related data are available at http://geneticassociationdb.nih.gov. Also, the association between diseases and genes can be obtained from the disease–metabolic subpathway network (DMSPN) [25].

3.2.5. *Protein interaction network*

The protein interaction network can be indirectly used to resolve this issue. The protein–protein interaction (PPI) data were derived from the Human Protein Reference Database which created proteomic information pertaining to human proteins [36]. To achieve a satisfactory performance, some researchers also employ such a network. Protein–disease associations can be obtained from the DISEASES database at http://diseases.jensenlab.org.

3.3. Computational Methods to Predict Disease-Related miRNA

To the best of our knowledge, this chapter is the first to review the state-of-the-art methods. Generally speaking, there are two main types of methods to address disease–miRNA prediction. One is experimental identification, which encounters many bottlenecks (e.g. it is time-consuming and expensive). Herein, computational means, proposed to supplement the experimental method, drastically reduce the number of candidate miRNAs. All these computational methods have one consistent purpose: predicting the relationship between the disease and miRNAs. For ease of understanding, we state our issue intuitively in Fig. 3.1. Some verified miRNA–disease associations are shown as straight lines, while potential relationships are shown as dotted lines. Disease and miRNA network data can be downloaded directly.

Our goal is to judge whether a miRNA is related to a specific disease. Of the massive methods, the key is how to define similarity scores (e.g. phenotype–miRNA similarity scores, miRNA similarity score, phenotype similarity score). Two biological assumptions are used to identify the associations between miRNAs and diseases.

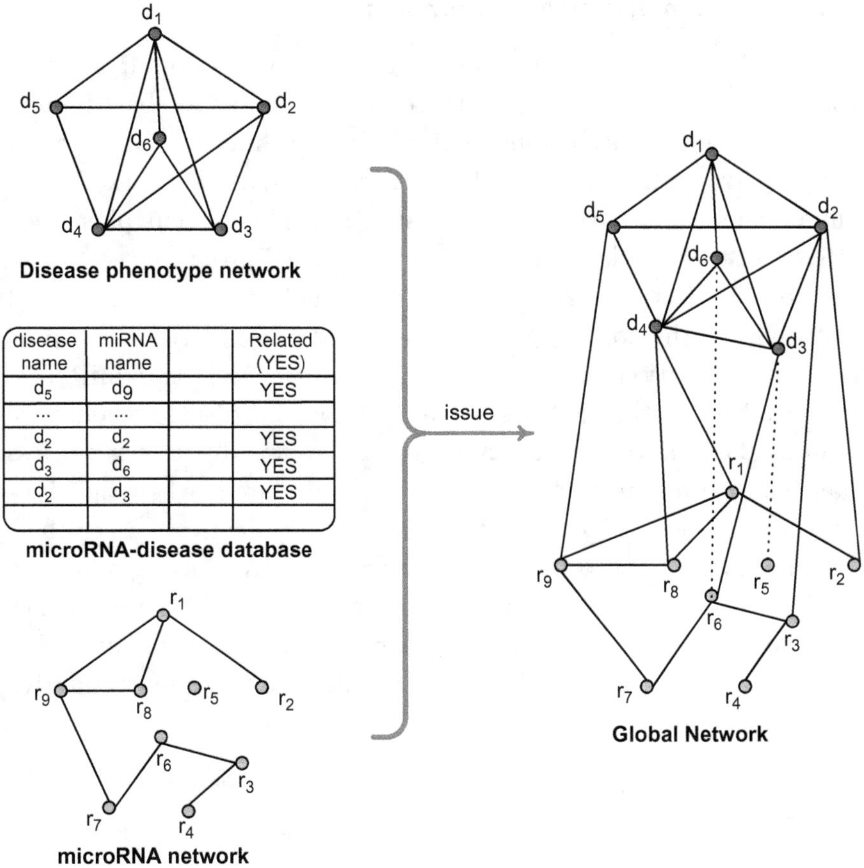

Fig. 3.1. Illustration of the network by using a special example.

Some researchers compute the similarity score based on the tendency that miRNAs are associated with phenotypically similar diseases. However, similarity score cannot be computed for diseases that have no known associated miRNAs, whereas methods based on the theory that functionally related miRNAs likely tend to be associated with phenotypically similar diseases can work for all diseases [20]. Here, we introduce, in chronological order, the proposed approaches from the following two aspects: similarity measure approaches (see Section 3.3.1) and machine learning means (see Section 3.3.2).

3.3.1. *Similarity measure methods*

In 2009, Jiang *et al.* [19] proposed the first computational method, which focuses on how to prioritize disease miRNAs based on a human phenome–miRNAome network. In their method, Jiang *et al.* first constructed a functionally related miRNA network and a human phenome–miRNAome network [19]. The human phenome–miRNAome network data can be downloaded from the database described in Section 3.2. The miRNAome network was constructed according to the following two considerations: miRNA functional models information and shortest pathway between two miRNAs. Specifically, the threshold is set at 0.3 for the disease phenotype network data. If the similarity score of the two diseases is no less than 0.3, the diseases are considered to be related; when the similarity score is equal to 1, the two diseases are identical. After that, a scoring system is proposed to prioritize miRNA. Sorting all of the miRNAs based on this score, the top-ranked miRNA has a higher probability to associate with a specific disease. Herein, for a disease of interest, d, the miRNA score is calculated by the cumulative hypergeometric distribution:

$$\text{score} = 1 - \sum_{i=m}^{M} \frac{\binom{M}{i}\binom{N-M}{n-i}}{\binom{N}{n}}$$

where N is the total number of miRNAs in the entire miRNA similarity network data. Here M is the total number of miRNAs known to be related to disease d, n is the number of miRNAs in the corresponding miRNA module and m is the number of miRNAs associated with similar diseases and is found in the corresponding module. The scores, as calculated by the above formula, are shown in Fig. 3.2.

For the method of Jiang *et al.* [19], the area under the ROC curve (AUC) is 0.758, indicating that this method achieved reasonable performance in identifying known disease-related miRNAs. This performance can be improved by later methods.

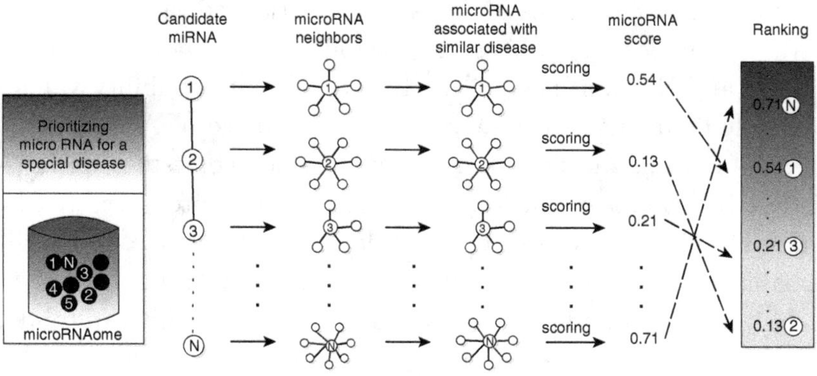

Fig. 3.2. Flow path of algorithm.

In 2010, Jiang *et al.* [20] proposed a new method, based on genomic data integration, to solve the above problem, to some extent. The Naïve Bayes model was used to integrate multiple types of data resources and build a model to predict functional associations between genes. The associations between a disease and gene are denoted as vectors, V_d, while the associations between the miRNA and target genes are denoted as vectors, V_m. For a given disease, Jiang *et al.* computed the similarity score for every miRNA and ranked the scores from highest to lowest. There is a high probability that the top-ranked miRNA score will predict a disease-related miRNA.

Chen *et al.* [10] applied random walk on the miRNA–miRNA functional similarity network and achieved an AUC of 86.17%. The random walk method, successfully applied in prioritizing candidate disease genes, randomly simulated a walker's transition from its current nodes to neighbours in the network starting at some given seed nodes [39]. There are three steps as follows: First, decide the initial probability of each mircoRNA; second, execute random walk on a miRNA functional similarity network; third, obtain the stable probability of random walk and rank the candidate miRNAs. For a given disease, d, Shi *et al.* [39] denoted that miRNAs are confirmed to be associated with the disease as seed nodes of random walk, and all the remaining miRNAs are candidates. They initialized (a) the probability value of nodes $p(0)$, where the possibility value of every seed node is equal and the sum is equal to 1, and (b) the possibility value of every candidate node whose value is equal to zero. A matrix, W, is available from the previous database, which consists of miRNA–miRNA functional similarity data. Here, in every iteration k, Chen *et al.* [39] allowed the restart of random walk at source nodes with probabilities $p(k)$, as a vector, representing the probability of the random walk at a node i at step k. Hence, the probability of random walk is defined as

$$p(k+1) = (1-r)W_{P(k)} + rp(0)$$

The condition for terminating the iterations is that the number representing the change between $p(k)$ and $p(k+1)$ is less than the cutoff value of 10.6. After the completion of the iterations, this method obtains a stable probability, $p(\infty)$. Based on the value of $p(\infty)$, Chen *et al.* ranked the candidate miRNAs related to the disease of interest, d. The higher the score of a miRNA, the higher the probability that it is associated with a known disease, d. The RWRMDA algorithm, visualized in Fig. 3.3, was tested by the leave-one-out cross-validation and had a satisfying performance compared with previous methods.

In 2013, Chen and Zhang [6] focused on computing a similarity and proposed a similarity-based method. Three strategies — (miRNA-based similarity inference) (MBSI), (phenotype-based similarity inference) (PBSI) and (network-consistency-based inference) (NetCBI) — based on different data networks, achieved AUC values of 74.83%, 54.02% and 80.66%, respectively [48]. To aid in the understanding of the above strategies, Chen *et al.* [48] introduced the following terminologies: (1) two sets, $M = \{m_1, m_2, \ldots, m_n\}$ and $P = \{p_1, p_2, \ldots, p_m\}$, to represent miRNAs and phenotypes; (2) a bipartite graph G (M, P, E), where $E = \{e_{ij} : m_i \in M,\ p_i \in P\}$. If a miRNA, i, has a relationship with a phenotype, j, an edge e_{ij} is drawn in graph G. From the perspective of method implementation, we stored the graph, G, in an $m \times n$ adjacent matrix,

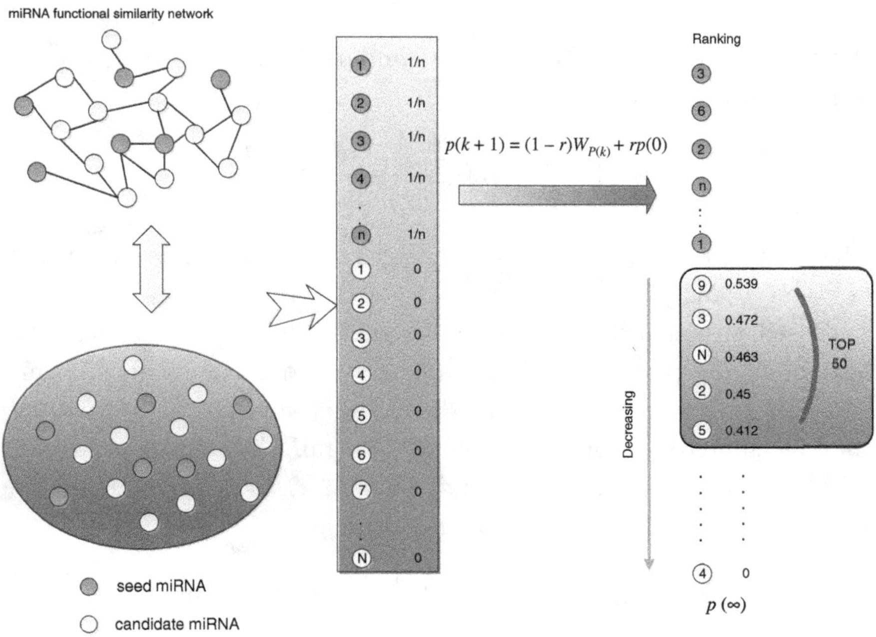

Fig. 3.3. Flowchart of RWRMDA.

$A = \{a_{ij}\ (i<n, j<m)\}$, where $a_{ij} = 1$ if a miRNA, i, and a phenotype, j, are verified as related; otherwise, $a_{ij} = 0$; (3) to store the data, three adjacency matrixes $A(m \times n)$, $M(n \times n)$ and $P(m \times m)$, which can be downloaded from the above databases.

In the MBSI method, Chen *et al.* proposed that for a given disease, d, if a miRNA, i, is related to the disease, d, other miRNAs, which are similar with the miRNA, i, have high probabilities to regulate the disease, d. Based on this consideration, they computed a similarity score between m_i and p_i using the following MBSIscore formula:

$$\text{MBSIscore}_{ij} = \frac{\sum_{l=1,l\neq i}^{n} S(m_i, m_l)a_{lj}}{\sum_{l=1,l\neq i}^{n} S(m_i, m_l)}$$

where $S(m_i, m_l) = M_{il}$ and M_{il} is a miRNA functional similarity value between m_i and m_l.

For the same consideration, they proposed that a miRNA, i, had a high probability to associate with the disease, which is similar to a disease, d, known to be linked to the miRNA, i. Herein, they used the PBSIscore formula to obtain a similarity score between m_i and p_i as follows:

$$\text{PBSIscore}_{ij} = \frac{\sum_{l=1,l\neq i}^{n} S(p_i, p_l)a_{lj}}{\sum_{l=1,l\neq i}^{n} S(p_i, p_l)}$$

where $S(p_{i,}p_l) = P_{il}$ and P_{il} is a miRNA functional similarity value between p_i and p_l.

Based on network consistency, to gain better performance, they integrated the phenotype similarity network data and miRNA similarity network data. Based on the graph Laplacian of $M(n \times n)$, they calculated the similarity scores of a miRNA i linked to other miRNAs. Normalizing $M = M(:, i)/\text{sum}(M(:, I))$, a vector m of the graph Laplacian is derived from the formula

$$\min_{\tilde{m}} \sum_{i,j} \bar{M}_{i,j}(\tilde{m}_i - \tilde{m}_j)^2 + \frac{1-a}{a} \sum_{i} (\tilde{m}_i - m_i)^2$$

To reduce unnecessary calculations, they found that a close solution to this formula is $\tilde{m} = (1-a)(I - a\bar{M})^{-1}$. Similarly, this graph Laplacian score is also applied to compute a phenotype vector, $\tilde{p} = (1-\beta)(I - \beta\bar{P})^{-1}$, where $\bar{P}$ is the normalization of P and $\alpha, \beta \in (0, 1)$ are parameters. A similarity score between a miRNA, i, with a phenotype, j, was computed via the Pearson correlation coefficient score by the following equation:

$$\text{NetCBIscore}(\tilde{m}, \tilde{p}, a) = \text{corr}(a\tilde{p}, \tilde{m})$$

where a is a vector from the matrix $A(m \times n)$. A phenotype is more likely to link with a miRNA, whose NetCBIscore is the higher one.

In 2013, based on the RWR algorithm, an improved method was proposed by Shi *et al.* [39]. In identifying known cancer-related miRNAs for the nine human cancers, a complementary network, containing miRNA–target networks, disease–gene networks and protein–protein networks achieved a satisfactory performance with an AUC ranging from 71.3% to 91.3% [39]. Based on previous biological knowledge that miRNAs regulated diseases through their target

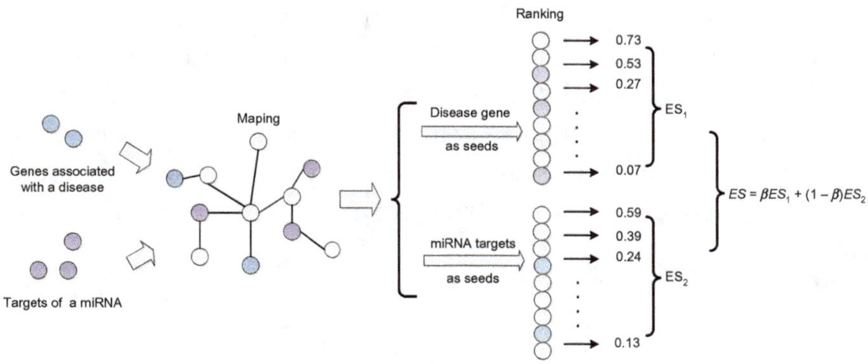

Fig. 3.4. Calculating similarity score via mapping PPI network.

genes, the hypothesis was put forth that if a miRNA target gene is associated with a disease gene, there is a high probability that the miRNA will affect the disease. Specifically, Shi *et al.* mapped disease genes and miRNA target genes on the PPI network and obtained two ranked lists of genes, derived from the RWR algorithm with different seeds. As a part of their method, Shi *et al.* performed the following three-step process — Step 1: set the disease genes as seeds; Step 2: on the seeds in step 1, do the gene set enrichment analysis (GSEA) to determine the ES score; Step 3: set the miRNA genes as seeds and repeat step 2 for these seeds. The ES score (the sum of ES1 and ES2) is computed by the following equations and visualized in Fig. 3.4:

$$ES_1 = \max\left(\sum_{g_j \in TG, j \leq i} \sqrt{(N-n_1)/n_1} - \sum_{g_j \in TG, j \leq i} \sqrt{n_1/(N-n_1)}\right)$$

$$ES_2 = \max\left(\sum_{g_j \in TG, j \leq i} \sqrt{(N-n_2)/n_2} - \sum_{g_j \in TG, j \leq i} \sqrt{n_2/(N-n_2)}\right)$$

$$ES = \beta ES_1 + (1-\beta) ES_2$$

Shi *et al.* used the p-value to measure the significance of an association between a miRNA and a disease and set the threshold to determine whether a miRNA is related to a given disease, d, without ranking miRNAs.

Xuan *et al.* [48] proposed another novel method, HDMP, based on the weighted k most similar neighbours to predict disease-related miRNAs [48]. For a miRNA, u, and a miRNA, v, the similarity of $\mathrm{Misim}\,(u,v)$ was computed. Xuan *et al.* denoted two disease sets, DT_u and DT_v, which contain all diseases related to miRNA u and v, respectively. Using the following equation, a $\mathrm{Misim}(u,v)$ value is calculated as

$$\mathrm{Misim}(u,v) = \frac{\sum_{1\leq i\leq |DT_u|} S(d_i, DT_v) + \sum_{1\leq j\leq |DT_u|} S(d_j, DT_u)}{|DT_u| + |DT_v|}$$

where $|DT_u|$ is the number of diseases in the set DT_u and $S(d_i, DT_v)$ is the similarity score between a disease d_i and the diseases in set DT_v. (For greater detail, see the flow chart in Fig. 3.5.) To obtain the value of $S(d_i, DT_v)$, Xuan *et al.* denoted two kinds of similarities between the two diseases: (1) the semantic similarity score $\mathrm{SS}(d_i, d_j)$ and (2) the phenotype similarity $\mathrm{PS}(d_i, d_j)$. Specifically, these similarities are computed based on the fact that diseases from the same level have a different contribution value because of the different frequency of occurrence of diseases in the GAD. Incorporating the semantic similarity and the phenotype similarity, the similarity between d_i and d_j is defined as $DS(d_i, d_j)$. Herein, Xuan *et al.* first denoted the similarity scores between a disease, d_1, and each of the diseases within a set of diseases, DT_v. Then, from among these similarity scores, they determine the maximum similarity score according to the following formula:

$$S(d_1, DT_v) = \max(\mathrm{SS}(d_1, d_{j_1}), \mathrm{SS}(d_1, d_{j2}), \ldots, \mathrm{SS}(d_1, d_{j_n}))$$

The method of Xuan *et al.* computes the similarity score between two diseases by Directed Acyclic Graph (DAG), which shows the relationship of diseases in different levels. For different levels, they denoted a different contribution factor. In the zeroth level, there is only the disease itself, so they set the factor at 1. As the levels increase, the contribution factor decreases exponentially. For example, the contribution factor in the ith level is Δ_i, $(\Delta = 0.5)$. (For an

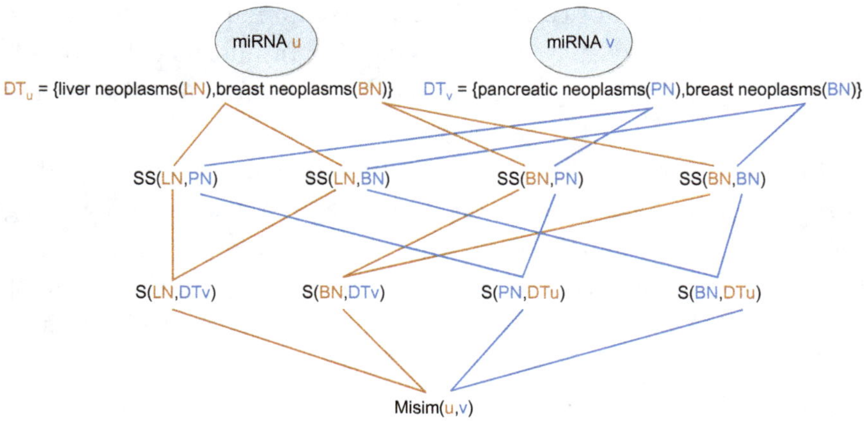

Fig. 3.5. Calculating miRNA functional similarity.

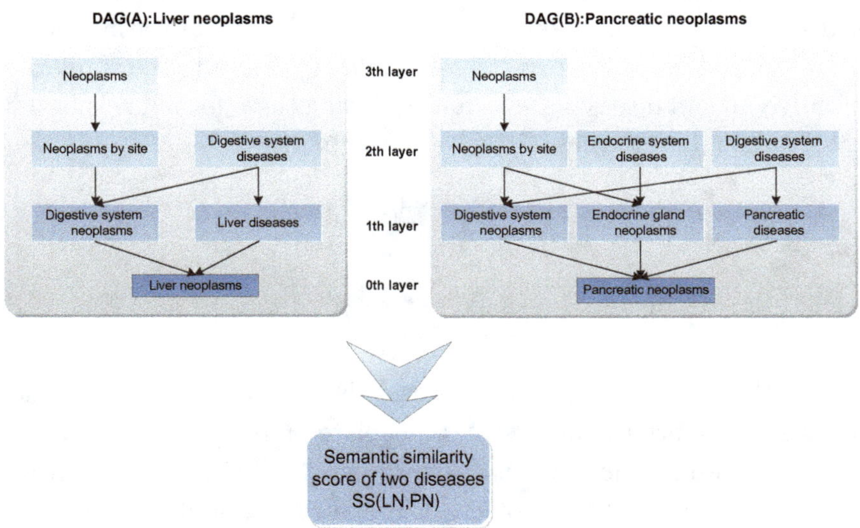

Fig. 3.6. Calculating disease's similarity via DAG.

example of DAG, see Fig. 3.6.) Considering two diseases, if a partition in one is the same as a partition in the other, then a value contribution is given to the similarity score. Many associations exist between the diseases, and all the associations also provide a value contribution to the similarity score. Through the above description, the similarity scores of two miRNAs are computed and a symmetric functional similarity matrix is constructed. In addition, by considering the miRNA family and the cluster information, this method attains a satisfactory performance.

In 2013, Mørk *et al.* [31] proposed another protein-driven inference method, which predicts miRNA–disease associations by using the linkage among miRNAs, proteins and diseases [31]. Specifically, for miRNA–disease associations, they downloaded miRNA–target data from the previous database and then mapped all miRNAs to miRBase identifiers and all targets to the ensemble protein identifiers using the STRING aliases file. Every association between a miRNA, m, and a protein, p, has a similarity score $T(m,p)$. For protein–disease associations, they downloaded the complete dataset from the DISEASES database and obtained a similarity score $Z(d,p)$ between a disease, d, and a protein, p. The mapped data are available at http://mirpd.jensenlab.org. Two formulae to compute the similarity of miRNAs and diseases are as follows: using these formulae is better than randomly selecting miRNA–disease pairs.

$$\text{Score}_1 = \sum_{P^* \in P_M \cap P_D} T(m,p^*)Z(d,p^*)$$

$$\text{Score}_2 = \max_{P^* \in P_M \cap P_D} T(m,p^*)Z(d,p^*)$$

Unlike the methods mentioned above, the method of Xu *et al.* [46] focuses on the prioritization of cancer-related miRNAs by integrating phenotype associations of multiple diseases with matched miRNA and mRNA expression profiles. With their method, the average AUC scores for 11 cancers were 75.84% and 73.59% based on the miR2Disease and HMDD databases, respectively. Their method computed results without using any prior information about whether miRNAs are related to a given disease. Based on the biological assumption, this method judged the relationship of a miRNA and a disease via computing the functional similarity score of miRNA–targets and disease genes from the following three perspectives: (1) the subontology Biological Processes (BP) of the Gene Ontology

(GO), (2) the Kyoto Encyclopedia of Genes and Genomes (KEGG) and (3) the topological structure of the Protein–Protein Interaction Network (PPIN) [17]. From a different viewpoint, they calculated an overlap score to store similarity scores between miRNA targets and disease genes for a specific disease. Then, they achieved the final functional similarity score by integrating the three overlapping scores. Based on the final score, potential miRNA associations were prioritized for a given cancer type.

In MIDP [49], Xuan *et al.* improved the random walk method and obtained a better prediction performance. For each given disease, they considered miRNAs differently and divided miRNAs into two parts. One defined as "labelled miRNA" that have known relationships with the given disease. Other defined as "unlabelled miRNA" that have no known relationships with the given disease. Then, they divided the miRNA similar matrix into two parts (labelled miRNA matrix and unlabelled miRNA matrix). After that, the random walk method was applied in two miRNA similar networks and the result scores from these two parts were added as the final similar score, which can be used to decide which miRNA is highly related with a given disease.

In 2016, Zeng *et al.* [52] applied two methods based on the multipath and achieved an area under the ROC curve of 0.981 and 0.984. One is the (HeteSim_multipath) (HSMP) and other is the HeteSim_SVM (HSSVM). HeteSim is a path-constrained measure, which can be used to calculate the relatedness of objects in heterogeneous networks. First, they define the search path including two type nodes (phenotypes and miRNAs) as $A_1 \to A_2 \to \cdots \to An$, where $A_1, A_2, \ldots$ belong to phenotype or miRNA. Second, based on a constrained path, they divided a path in two parts ($\text{Path}_{\text{left}}$ and $\text{Path}_{\text{right}}$). Specifically, they defined the length of the path as L. If L is even, $\text{Path}_{\text{left}} = A_1 \cdots A_{\text{mid}-1}A_{\text{mid}}$ and $\text{Path}_{\text{right}} = A_{\text{mid}}A_{\text{mid}+1} \cdots A_{L+1}$, simply, where mid $= L/2 + 1$. While, if the length L is odd, we divide the path twice by setting mid $= (L+1)/2$ and mid $= (L+3)/2$, respectively, and then compute the average score as the final HeteSim Score. In the following, for each of the two divisions, we obtain a HeteSim score and the final score is the average of the two scores. The HeterSim score can be calculated by the following formula:

$$\text{HeteSim}(a, b|\text{Path}) = \frac{PM_{\text{left}}(a, :)PM_{\text{right}}(:, b)}{\sqrt{\|PM_{\text{left}}(a, :)\| \|PM_{\text{right}}(:, b)\|}}$$

Third, they used the following strategy to consider the idea that different paths contribute differently because of the length of paths. They assumed a constant β, where β is used for dampen contributions of longer paths.

$$S_{A,B}^{\text{HSMP}}(a, b) = \sum_{l=2}^{\infty} \sum_{i=1}^{|\text{paths}_{A,B}^{\{L\}}|} (\beta^{l-1} * \text{HeterSim}(a, b|\text{paths}_{A,B}^{\{L\}}(i)))$$

where A, B represented a miRNA or a disease node and l is for the length of path.

Another method, HSSVM, proposed by Zeng *et al.* distinguished with the method HSMP in the third step. To learn the contributions of different paths by Support Vector Machine (SVM) is the third step in the HSSVM.

Chen *et al.* proposed two methods (MBSMDA [9] in 2015 and PBMDA [51] in 2017) based on the network to predict potential associations between the miRNAs and diseases. These two methods are based on the same heterogeneous networks and they added the miRNA and disease kernel similarities to reconstruct networks. For every miRNA and disease, they computed a Gaussian interaction profile kernel similarity. First, they defined a binary vector $\text{IP}(d(i))$ as the interaction profile of disease $d(i)$ by observing whether disease $d(i)$ is connected with each miRNA or not. Then the Gaussian interaction profile kernel similarity between $d(i)$ and $d(j)$ was computed by the following equation:

$$KD(d(i), d(j)) = \exp(-\gamma_d \text{IP}(d(i)) - \text{IP}(d(j))^2)$$

where γ_d is a parameter used to control the kernel bandwidth. Similarly, they computed the kernel similarity for every miRNA in the same way.

In Within and Between Score for MiRNA–Disease Association (WBSMDA) [9] methods, Chen *et al.* obtained a final similar score from two parts (within score and between score). The within score is defined to find the miRNA score with the investigated miRNA among the group of miRNAs with known association with the investigated disease, while the between score is defined to find the miRNA score with the investigated miRNA in the group of miRNAs without known association with the investigated disease.

In Path-Based MiRNA–Disease Association (PBMDA) [51] methods based on demonstrated networks, they defined a path as a connection between a miRNA and a disease and used a specific depth-first search algorithm to search all paths which are not allowed to include more than three paths. Meanwhile, they defined an exponential decay function, $F(p)$, to control the contribution of different path lengths. Then, they obtained the final score by the following formula:

$$\mathrm{score}(m_i, d_j) = \sum_{w=1}^{n} \left(\prod P_w\right)^{F(p)}$$

where n is the number of paths.

Similarly, another method based on the heterogeneous networks is the network consistency method.

In the method, Gu *et al.* [15] reconstructed the heterogeneous network and computed the miRNA–disease similarity score in two space (miRNA space and disease space). Finally, the miRNA space projection score and disease space projection score are combined and normalized as the miRNA–disease projection score.

3.3.2. *Machine learning methods*

In 2010, Xu *et al.* [47] applied the machine learning method to predict the relationship between miRNAs and diseases. This approach aims to distinguish positive miRNA–disease associations from large-scale negative miRNA–disease associations. The primary focus of this method is to extract features from the miRNA–disease network data

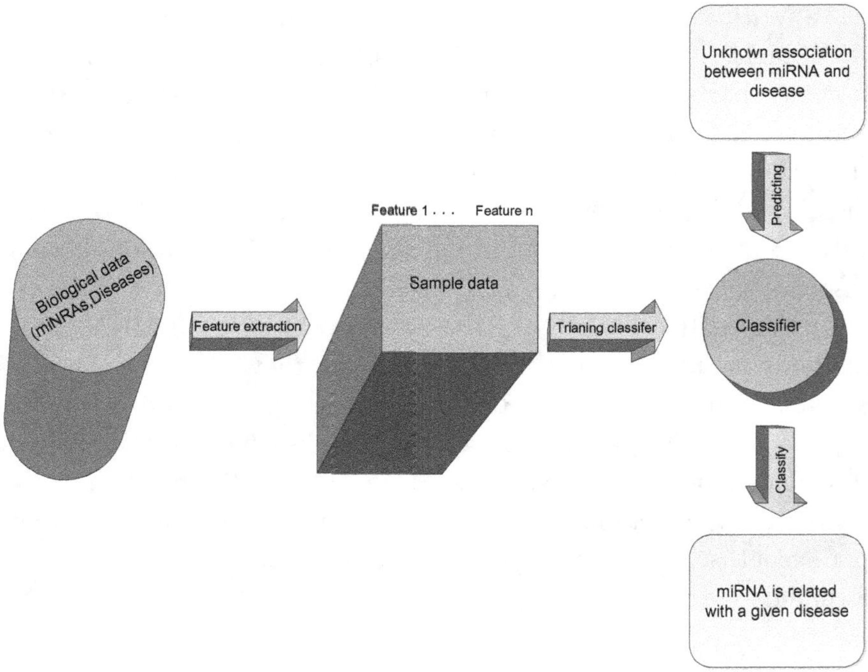

Fig. 3.7. Flowchart of machine learning methods.

and train a Support Vector Machine (SVM) classifier, which obtained the area under the Receiver Operating Characteristic (ROC) curve of up to 0.8872 in cross-validation tests [47]. Jiang *et al.* [18] extracted different features to obtain an AUC value of 0.8884 in ten-fold cross-validation [18]. Figure 3.7 shows the flowchart of machine learning approaches. First, they constructed two feature vectors consisting of two feature sets. One feature set is primarily related to miRNA information; whereas, the other is disease phenotype information.

Feature set primarily related to miRNA information: For the first feature set, each element value was calculated via the functional similarity score between a miRNA and the other miRNAs related to a given disease. To compute the functional similarity score between the two miRNAs, the miRNA interaction data were downloaded from the aforementioned database and denoted as $\vec{V}_m = \{w_{m_i,g_1}, w_{m_i,g_2}, w_{m_i,g_3}, \ldots\}_{(i=1,2)}$, where $w_{mi,gj}$ is the interaction score predicted by the PITA [22] prediction algorithm. In this definition, the interaction scores of all genes in the human genome with miRNA, i, are used. The interaction score of the non-interacting miRNA–target is set to zero. Thereafter, they obtained a miRNA pair functional similarity score that is a Pearson's correlation coefficient between $\vec{V}_{m_1}$ and $\vec{V}_{m_2}$. A schematic illustration of the machine learning method is shown in Fig. 3.7.

Disease phenotype information: In the second feature set, each element value was a disease phenotype similarity score, available from MimMiner, which was introduced in the database partition. If the score in the MimMiner was less than the noise threshold of 0.3, then we set the similarity score of the two diseases as zero. After the aforementioned operations, the feature vector of each miRNA–disease association contained 171 dimensions [10].

An n-fold cross-validation procedure was then performed. The positive and negative samples were randomly distributed into n parts. The $n-1$ parts of the samples were selected as the training set, and one sample part was the testing part; the selection was repeated 10 times to ensure that each of the n parts was considered as the testing part. Finally, the authors used the trained SVM classifier to predict the potential miRNA–disease relationships and classify the unlabelled data.

In 2014, considering the limitations in previous methods, Chen and Yan [7] proposed Regularized Least Squares for miRNA–Disease Association (RLSMDA). RLSMDA was designed to construct a continuous classification function, which reflects the probability that each miRNA is related to a given disease. This function works for diseases without known related miRNAs. This approach is a semi-supervised and global method, which simultaneously prioritizes associations for all the diseases without the negative samples. As successfully confirmed by experiments, this approach also correctly predicts the potential miRNA among the top three of 32 diseases and 34 disease–miRNA associations. In this method, the researchers used the disease data from a method similar to the HDMP method. To solve the problem of no negative samples, a semi-supervised classifier was constructed under the framework of Regularized Least Squares, obtained by defining and minimizing a cost function. This cost function was trained in the miRNA and disease networks, and the forward strategy was then used to optimize the classification function. Finally, researchers obtained two different optimal classifiers in both miRNA and disease spaces. Both spaces simultaneously contributed to solve this problem, and a parameter was set to balance the final scores. A high score leads to a high probability of association between a miRNA and a given disease.

In 2015, Chen *et al.* [8] proposed a method based on the restricted Boltzmann machine for multiple types of miRNA–disease association prediction. According to their evidences, they divided the associations into four catalogues (miRNA–target associations, circulation associations, epigenetics associations and genetics associations). First, they defined four binary vectors for each associations between miRNAs and diseases. Second, for each miRNA, an RBM was constructed. There are visible units that represent diseases and hidden units that represent unknown features of the associations in the RBM. Third, they used the RBM model to capture associations' type by updating hidden units and visible units iteratively. Finally, they computed conditional probabilities of link generation based on RBM outputs.

In 2016, Pasquier and Gardès *et al.* [34] proposed a new method with the vector space model to predict associations between miRNAs and diseases. All their work is based on the hypothesis that miRNA–disease associations can be revealed by various semantics, so they integrate data as much as possible to define the associations in a high-dimensional vector space. The data resources they used involved miRNA–target information, miRNA–family information, miRNA–disease information, miRNA–word information and miRNA–neighbour information. They used all of this information to construct a large matrix X. After that, they applied the Singular Value Decomposition (SVD) method to conduct dimensionality reduction. By using SVD, the original matrix X is divided into three matrix parts as shown below.

$$X = UAV^T$$

Through mediating the dimension of the matrix, A, they obtained a better prediction performance.

3.3.3. *Comparisons with different approaches*

In this section, we review the different computational methods to predict the disease–miRNA associations.

Table 3.1 provides an extensive overview of the methods that have been developed to date. In this table, the type of networks and databases used in the approaches and other fundamental information are compared. To understand the research progress, the same categories are discussed in a roughly chronological order.

Table 3.1 presents an overview of the 11 approaches, which are continuously being developed and improved. This table also indicates that the prediction performance tested with the ROC curve increased from 75.80% to 84.50%.

By applying the previous methods, the authors simply used the local network and limited information. Integrating more biological information networks guides researchers in considering the complete data and obtaining a satisfactory prediction performance. The method of Jiang *et al.* [19] could be improved by considering the following: First, the miRNA interaction network constructed using this method merely considered two aspects. More related information to build the miRNA network results in a more reliable miRNA association that predicts the disease-related miRNAs. Second, the Boolean network was utilized to store the phenotype–phenotype associations. Changing the Boolean network into a weight network is an improvement that can lead to more detailed information. By replacing the Boolean network with weight network, Jiang *et al.* [19] improved the prediction performance of their study. Köhler *et al.* [53], Shi *et al.* [39] and Xuan *et al.* [48] obtained a high AUC value, but the performance

Table 3.1. Comparison of disease related microRNA prediction methods. M–D, microRNA–disease associations; M–M, microRNA–microRNA associations; D–D, disease phenotype–disease phenotype associations; D–G, disease–gene associations; M–G, microRNA–target gene associations; D–P, disease–protein associations; M–P, microRNA–protein associations; P–P, protein–protein associations; EP, expression profiles.

Method	Type of networks	Type of databases	Amount of data	AUC values	Type of test	Time	References
Jiang *et al.*	M–D M–M D–D	HMDD, miR2Diseas PITA, TargetScan, miRBase MimMiner	270(M–D assos) 145872; 205587 (M–G assos)	75.80%	Cross-fold validation	2009.8	[19]
Jiang *et al.*	M–D M–M P–P	miR2Diseas PITA PPI	62 microRNAS 145872(M–G assos) 3106928(P–P assos)	—	Ranking	2010	[20]
Chen *et al.*	M–D M–M	HMDD, miR2Diseas, dbDEMC misim	1616(M–D assos)	86.17%	Leave-one-out cross-validation	2012.8	[10]
Chen and Zhang	M–D M–M D–D	HMDD, miR2Diseas misim MimMiner	242(M:99, D:51)	80.66%(NetCBI) 74.83%(MBSI) 54.2%(PBSI)	Leave-one-out cross-validation	2013.6	[6]
Shi *et al.*	D–G M–G PPI	DMSPN miRanda, PicTar, TargetScan, DIANA-microT, RNA22, RNAhybrid, miRBase Targets PPI	15149(D:412, G:2831) 52828(M:566, g:8085) 36867(G–G assos, G:9453)	71.3%–91.3% (in nine human cancers)	Cross-validation	2013	[39]

Xuan *et al.*	*M–D* *D–D*	HMDD MimMiner	4739(*M*:474, *D*:268) (*D*:5080)	82.47% (average in 18 diseases)	Five-fold cross-validation	2013.8	[48]
Mørk *et al.*	*M–G* *D–P*	miRanda, TargetScan DISEASES	14599 (*M*:1169 *P*:1570 *D*:738)	—	Ranking	2013.11	[31]
Xu *et al.*	*M–D* *M–G* *D–D*	HMDD, miR2Diseas miRanda, PicTar, TargetScan, DIANA–microT, RNA22, RNAhybrid, miRBase Targets MimMiner	52828(*M*:566, *G*:8085)	75.84% (average)	Cross-validation	2014.7	[46]
Xuan *et al.*	*M–D* *M–G* *D–D*	HMDD misim DAG	5100(*M*:490, *D*:326)	78.6%–94.5% (in nine human cancers)	Cross-validation	2015	[49]
Zeng *et al.*	*M–D* *M–M* *D–D*	HMDD misim MimMiner	242(*M*:271, *D*:5080) 56289(*M*:271) 20285172(*D*:5080)	98.1% (average)	Leave-one-out cross-validation	2015	[52]
Chen *et al.*	*M–D* *M–M* *D–D*	HMDD misim DAG	—	80.31% (average)	Leave-one-out cross-validation	2016.6	[9]
You *et al.*	*M–D* *M–M* *D–D*	HMDD misim DAG	—	83.41% (average)	Leave-one-out cross-validation	2017	[51]

(*Continued*)

Table 3.1. (*Continued*)

Method	Type of networks	Type of databases	Amount of data	AUC values	Type of test	Time	References
Gu *et al.*	M–D M–M D–D	HMDD misim DAG	1395(M:271, D:137)	89.58%	Leave-one-out cross-validation	2016.10	[15]
Xu *et al.*	**EP** M–G	**miRanda, PicTar, TargetScan, DIANA–microT,**	**484043(M:320, G:12713)**	**88.72%**	**Cross-validation**	**2011.7**	[47]
Jiang *et al.*	M–D M–G D–D	**HMDD, miR2Diseas** PITA, TargetScan, miRBase MimMiner	**270(D:53, M:118)** 145872; 205587(M–G assos)	**88.84%**	**Ten-fold** cross-validation	**2013**	[18]
Chen and Yan	M–D M–M D–D	HMDD misim DAG	1616(M–D assos)	84.50%	Leave-one-out cross-validation	2014.6	[7]
Chen *et al.*	M–D	HMDD	—	86.06%	Leave-one-out cross-validation	2015.9	[8]
Pasquier and Gardès	M–D M–M D–D	HMDD misim DAG, MeSH	10737(M:559, D:369)	86.7%	Five-fold cross-validation	2016.7	[34]

of their methods depended on the selected parameters, including the restart probability, p-value and number of neighbours (k). Xuan *et al.* [48] utilized the similarity of the disease sets to compute the miRNA similarity score. Obtaining a specific miRNA similarity score is significant and results in an outstanding prediction performance. Several approaches, which have been applied by Köhler *et al.* [53] and Xuan *et al.* [48], could not predict diseases without the specified miRNAs. This problem was solved by utilizing all the machine learning methods [18, 47, 54] and any of the non-machine learning methods of Jiang *et al.* [18], Chen and Zhang [6] and Xu *et al.* [47]. However, a few researchers regard the unlabelled samples as negative in the application of machine learning methods, which may lead to a low prediction performance. Akyildiz and Chen [54] recently addressed this issue and obtained a better performance compared with the Random Walk with Restart for miRNA–Disease Association techniques [10] and high-dose methylprednisolone [48].

3.4. Conclusion

Previous biological theories indicate that researchers can use computational methods to supplement biological experimental methods to predict the potential associations between miRNAs and diseases. Considerable biological interaction information is represented as networks, such as miRNA–disease, disease phenotype, miRNA association, gene interaction and protein interaction networks. The data of these networks can be downloaded directly. Integrating the network data to identify the associations between miRNAs and diseases is ideal and helps researchers obtain a better performance. Moreover, society can completely understand a biological system. One key issue in integrating a network is computing the similarities of the two nodes, particularly between a miRNA and a disease. In this chapter, we enumerate all types of approaches to address the similarity measure, and elucidate their respective advantages and disadvantages. In general, the similarity score can be determined by considering the neighbours, shortest pathway and familial information. Machine learning methods do not directly measure the similarity score, but they consider such information through feature extraction.

In this chapter, we introduced several approaches to address the aforementioned issue and a few emerging problems that impeded the performance improvement of predictions. Future studies can focus on the following three aspects to improve the effects of forecasting performance. First, the miRNA interaction networks are not reliable because they were constructed using miRNA–target prediction tools, which contain numerous incorrect data. Hence, future scholars may develop more comprehensive data networks by using substantially related biological information. Second, future scholars may utilize updated data to obtain a better prediction performance. Finally, future researchers must consider more biological knowledge to predict the association between miRNAs and diseases. Comprehensive and important information lead to accurate predictions.

Acknowledgments

This work was supported by the Natural Science Foundation of China [Grant Numbers 61370010 and 61371179], the Natural Science Foundation of Fujian Province of China [Grant Number 2014J01253], and the Open Funding of Shanghai Key Laboratory of Intelligent Information Processing, China [Grant Number IIPL-2014-004].

Bibliography

[1] Abrahante, J. E. *et al.* The *Caenorhabditis elegans* hunchback-like Gene lin-57/hbl-1 controls developmental time and is regulated by microRNAs. *Dev. Cell* **4**(5), 625–637, 2003.

[2] Ambros, V. The functions of animal microRNAs. *Nature* **431**(7006), 350–355, 2004.

[3] Bartel, D. P. MicroRNAs: Genomics, biogenesis, mechanism, and function. *Cell* **116**(2), 281–297, 2004.

[4] Becker, K. G. *et al.* The genetic association database. *Nat. Genet.* **36**(5), 431–432, 2004.

[5] Betel, D. *et al.* The microRNA.org resource: Targets and expression. *Nucleic Acids Res.* **36**(1), D149–D153, 2008.
[6] Chen, H. and Zhang, Z. Similarity-based methods for potential human microRNA–disease association prediction. *BMC Med. Genomics* **6**(1), 12, 2013.
[7] Chen, X. and Yan, G.-Y. Semi-supervised learning for potential human microRNA–disease associations inference. *Sci Rep.* **4**, 5501, 2014.
[8] Chen, X. *et al.* RBMMMDA: Predicting multiple types of disease–microRNA associations. *Sci. Rep.* **5**, 13877, 2015.
[9] Chen, X. *et al.* WBSMDA: Within and between score for miRNA–disease association prediction. *Sci. Rep.* **6**, 2016.
[10] Chen, X., Liu, M.-X. and Yan, G.-Y. RWRMDA: Predicting novel human microRNA–disease associations. *Mol. BioSyst.* **8**(10), 2792–2798, 2012.
[11] Goh, K.-I. *et al.* The human disease network. *Proc. Natl. Acad. Sci. USA* **104**(21), 8685–8690, 2007.
[12] Griffiths-Jones, S. *et al.* miRBase: microRNA sequences, targets and gene nomenclature. *Nucleic Acids Res.* **34**(1), D140–D144, 2006.
[13] Griffiths-Jones, S. *et al.* Rfam: An RNA family database. *Nucleic Acids Res.* **31**(1), 439–441, 2003.
[14] Grimson, A. *et al.* MicroRNA targeting specificity in mammals: determinants beyond seed pairing. *Mol. Cell* **27**(1), 91–105, 2007.
[15] Gu, C. *et al.* Network consistency projection for human miRNA–disease associations inference. *Sci. Rep.* **6**, 2016.
[16] Hamosh, A. *et al.* Online Mendelian Inheritance in Man (OMIM), a knowledgebase of human genes and genetic disorders. *Nucleic Acids Res.* **33**(1), D514–D517, 2005.
[17] Jiang, Q. *et al.* miR2Disease: A manually curated database for microRNA deregulation in human disease. *Nucleic Acids Res.* **37**(1), D98–D104, 2009.
[18] Jiang, Q. *et al.* Predicting human microRNA–disease associations based on support vector machine. *Int. J. Data Mining Bioinf.* **8**(3), 282–293, 2013.
[19] Jiang, Q. *et al.* Prioritization of disease microRNAs through a human phenome–microRNAome network. *BMC Syst. Biol.* **4**(1), S2, 2010.
[20] Jiang, Q., Wang, G. and Wang, Y. An approach for prioritizing disease-related microRNAs based on genomic data integration. In *IEEE 3rd International Conference on Biomedical Engineering and Informatics* (*BMEI*), 2010.
[21] Jopling, C. L. *et al.* Modulation of hepatitis C virus RNA abundance by a liver-specific microRNA. *Science* **309**(5740), 1577–1581, 2005.
[22] Kertesz, M. *et al.* The role of site accessibility in microRNA target recognition. *Nat. Genet.* **39**(10), 1278–1284, 2007.
[23] Krek, A. *et al.* Combinatorial microRNA target predictions. *Nat. Genet.* **37**(5), 495–500, 2005.
[24] Krüger, J. and Rehmsmeier, M. RNAhybrid: microRNA target prediction easy, fast and flexible. *Nucleic Acids Res.* **34**(2), W451–W454, 2006.
[25] Li, X. *et al.* The implications of relationships between human diseases and metabolic subpathways. *PLoS one* **6**(6), e21131, 2011.

[26] Li, Y. *et al.* HMDD v2.0: A database for experimentally supported human microRNA and disease associations. *Nucleic Acids Res.* **42**, D1070–D1074, 2013.
[27] Maragkakis, M. *et al.* DIANA-microT web server: Elucidating microRNA functions through target prediction. *Nucleic Acids Res.* **37**, W273–W276, 2009.
[28] Megraw, M. *et al.* miRGen: A database for the study of animal microRNA genomic organization and function. *Nucleic Acids Res.* **35**(1), D149–D155, 2007.
[29] Meister, G. and Tuschl, T. Mechanisms of gene silencing by double-stranded RNA. *Nature* **431**(7006), 343–349, 2004.
[30] Miranda, K. C. *et al.* A pattern-based method for the identification of MicroRNA binding sites and their corresponding heteroduplexes. *Cell* **126**(6), 1203–1217, 2006.
[31] Mørk, S. *et al.* Protein-driven inference of miRNA–disease associations. *Bioinformatics* **30**(3), 392–397, 2013.
[32] Moss, E. G., Lee, R. C. and Ambros, V. The cold shock domain protein LIN-28 controls developmental timing in *C. elegans* and is regulated by the lin-4 RNA. *Cell* **88**(5), 637–646, 1997.
[33] Nam, S. *et al.* miRGator: An integrated system for functional annotation of microRNAs. *Nucleic Acids Res.* **36**(1), D159–D164, 2008.
[34] Pasquier, C. and Gardès, J. Prediction of miRNA–disease associations with a vector space model. *Sci. Rep.* **6**, 2016.
[35] Peng, K. *et al.* The Disease and Gene Annotations (DGA): An annotation resource for human disease. *Nucleic Acids Res.* **41**, D553–D560, 2012.
[36] Prasad, T. K. *et al.* Human protein reference database—2009 update. *Nucleic Acids Res.* **37**(1), D767–D772, 2009.
[37] Reinhart, B. J. *et al.* The 21-nucleotide let-7 RNA regulates developmental timing in *Caenorhabditis elegans*. *Nature* **403**(6772), 901–906, 2000.
[38] Sethupathy, P., Corda, B. and Hatzigeorgiou, A. G. TarBase: A comprehensive database of experimentally supported animal microRNA targets. *RNA* **12**(2), 192–197, 2006.
[39] Shi, H. *et al.* Walking the interactome to identify human miRNA–disease associations through the functional link between miRNA targets and disease genes. *BMC Syst. Biol.* **7**(1), 101, 2013.
[40] Slack, F. J. *et al.* The lin-41 RBCC gene acts in the *C. elegans* heterochronic pathway between the let-7 regulatory RNA and the LIN-29 transcription factor. *Mol. Cell* **5**(4), 659–669, 2000.
[41] van Driel, M. A. *et al.* A text-mining analysis of the human phenome. *Eur. J. Hum. Genet.* **14**(5), 535–542, 2006.
[42] Vasudevan, S., Tong, Y. and Steitz, J. A. Switching from repression to activation: microRNAs can up-regulate translation. *Science* **318**(5858), 1931–1934, 2007.
[43] Wang, D. *et al.* Inferring the human microRNA functional similarity and functional network based on microRNA-associated diseases. *Bioinformatics* **26**(13), 1644–1650, 2010.

[44] Wightman, B., Ha, I. and Ruvkun, G. Posttranscriptional regulation of the heterochronic gene lin-14 by lin-4 mediates temporal pattern formation in C. elegans. *Cell* **75**(5), 855–862, 1993.

[45] Xie, B. *et al.* miRCancer: a microRNA–cancer association database constructed by text mining on literature. *Bioinformatics* **29**(5), 638–644, 2013.

[46] Xu, C. *et al.* Prioritizing candidate disease miRNAs by integrating phenotype associations of multiple diseases with matched miRNA and mRNA expression profiles. *Mol. Biosyst.* **10**(11), 2800–2809, 2014.

[47] Xu, J. *et al.* Prioritizing candidate disease miRNAs by topological features in the miRNA target–dysregulated network: Case study of prostate cancer. *Mol. Cancer Ther.* **10**(10), 1857–1866, 2011.

[48] Xuan, P. *et al.* Prediction of microRNAs associated with human diseases based on weighted k most similar neighbors. *PLoS one* **8**(8), e70204, 2013.

[49] Xuan, P. *et al.* Prediction of potential disease-associated microRNAs based on random walk. *Bioinformatics* **31**(11), 1805–1815, 2015.

[50] Yang, Z. *et al.* dbDEMC: A database of differentially expressed miRNAs in human cancers. *BMC Genomics* **11**(4), S5, 2010.

[51] You, Z.-H. *et al.* PBMDA: A novel and effective path-based computational model for miRNA–disease association prediction. *PLoS Comput. Biol.* **13**(3), e1005455, 2017.

[52] Zeng, X. *et al.* Prediction and validation of association between microRNAs and diseases by multipath methods. *Biochim. Biophys. Acta (BBA)— Gen. Subj.* **1860**(11), 2735–2739, 2016.

[53] Köhler, S. *et al.* Walking the interactome for prioritization of candidate disease genes. *Am. J. Hum. Genet.* **82**(4), 949–958, 2008.

[54] Akyildiz, I. F. and Chen, R. J. A new random walk model for PCS networks. *IEEE J. Sel. Areas Commun.* **18**(7), 1254–1260, 2000.

Chapter 4

Evolutionary Algorithm for Solving Complex Multiobjective Optimization Problems

Ye Tian* and Xingyi Zhang†

School of Computer Science and Technology
Anhui University, Hefei, China
**field910921@gmail.com*
†xyzhanghust@gmail.com

In real-world applications, there are many problems containing multiple conflicting objectives to be optimized, which are called multiobjective optimization problems (MOPs). Since the objectives in MOPs are multimodal, discontinuous, non-differentiable or even discrete, general mathematic or statistic approaches encounter difficulties in dealing with them. By contrast, many bio-inspired algorithms have shown their superiority in tackling MOPs during the last two decades. Among these metaheuristics, the evolutionary algorithm has been proved to be one of the most powerful approach. Nevertheless, the complex MOPs are still big challenges for the general evolutionary algorithm, which are characterized by many objectives or a large number of decision variables. This chapter introduces several techniques for enhancing the evolutionary algorithm in solving complex MOPs, with respect to improving the efficiency and effectiveness, respectively.

4.1. Introduction to Multiobjective Optimization

Multiobjective optimization problems (MOPs) are commonly seen in real-world applications, especially in the areas of engineering [10], biology [14] and economics [12]. MOPs also widely exist in many artificial intelligence applications such as data mining [9], pattern recognition [6], image processing [7] and artificial neural networks [11]. MOPs are characterized by two or more objectives to be optimized, where the objectives are usually in conflict mutually. An MOP can be formulated as

$$\begin{aligned} \min_{x} \quad & f(x) = (f_1(x), \ldots, f_M(x))^T \\ \text{s.t.} \quad & g_i(x) \leq 0, \quad i = 1, \ldots p \\ & h_j(x) = 0, \quad j = 1, \ldots q \end{aligned}$$

where $x = (x_1, x_2, \ldots, x_D)^T \in X \subseteq \mathbb{R}^D$ indicates one solution, and X is the *decision space*. $f : X \to Y \subseteq \mathbb{R}^M$ consists of M objectives and Y is the *objective space*. $g_i(x)$ are p inequality constraints, and $h_j(x)$ are q equality constraints. When the number of objectives M is larger than three, MOPs are also known as many-objective optimization problems (MaOPs).

Definition 4.1. Solution x is said to Pareto dominate solution y, denoted as $x \prec y$, if and only if

$$\begin{cases} f_i(x) \leq f_i(y), & \forall i = 1, 2, \ldots, D \\ f_j(x) < f_j(y), & \exists j = 1, 2, \ldots, D \end{cases}$$

Definition 4.2. Solution x is said to non-dominated with solution y, if and only if

$$\begin{cases} f_i(x) < f_i(y), & \exists i = 1, 2, \ldots, D \\ f_j(x) > f_j(y), & \exists j = 1, 2, \ldots, D, j \neq i \end{cases}$$

Definition 4.3. Solution x is called a non-dominated solution in solution set P, if and only if

$$\nexists y \in P : y \prec x$$

Definition 4.4. Solution x is called a Pareto optimal solution, if and only if

$$\nexists y \in X : y \prec x$$

Definition 4.5. Pareto set (PS) denotes the set of all the Pareto optimal solutions in the decision space X.

Definition 4.6. Pareto front (PF) denotes the image of the PS in the objective space Y.

Due to the conflicting nature among objectives, there is no single optimal solution that can make all the objectives to be optimal simultaneously; instead, the goal of solving an MOP is to find a set of non-dominated solutions which is close to the PF (good convergence), and distributed uniformly and widely (good diversity).

As one of the most popular bio-inspired algorithms, the evolutionary algorithm has been proved to be effective in solving MOPs, where a set of non-dominated solutions can be found in a single run due to its population-based nature. As a result, there has been a growing interest in applying the evolutionary algorithm to solve MOPs, and these algorithms are called multiobjective evolutionary algorithms (MOEAs) [27]. However, MOEAs are often criticized because of their low efficiency and effectiveness in solving complex MOPs, which have many objectives or a large number of decision variables.

4.2. Improving the Efficiency of Multiobjective Evolutionary Algorithms

The efficiency of most MOEAs is subject to the computational complexity of non-dominated sorting, which is an important operation used in selection strategies to sort the population according to their dominance relations. Taking the NSGA-II [4] as an example, non-dominated sorting may consume more than 70% of the total runtime. Therefore, it is necessary to improve the efficiency of MOEAs by decreasing the computational complexity of non-dominated sorting.

The original non-dominated sorting works as follows:

Step 1: $i = 1$.
Step 2: Move all the non-dominated solutions from P to F_i.
Step 3: If P is empty, stop; otherwise $i = i + 1$ and go to Step 2.

It is obvious that each solution in the population P is divided into a front F_i after non-dominated sorting. Then, the solutions in former fronts are considered to be better than those in latter fronts.

The above procedure of non-dominated sorting holds a quite high time complexity, which is $O(MN^3)$ with N denoting the population size and M denoting the number of objectives. Although it was decreased to $O(MN^2)$ by the fast non-dominated sort [4], non-dominated sorting is still inefficient when the population size or the number of objectives is large. In the following sections, several more efficient non-dominated sorting methods will be introduced.

4.2.1. *Efficient Non-dominated Sort*

4.2.1.1. *Framework of ENS*

Unlike the fast non-dominated sort that sorts the solutions front by front, the efficient non-dominated sort (ENS) [21] determines the front number of each solution in turn. Let front(p) be the front number of solution p and Q is the set of solutions in the population which dominate p, then we have

$$\text{front}(p) = 1 + \max_{q \in Q} \text{front}(q)$$

The fast non-dominated sort compares each solution with all the others, but the above formula implies that p just needs to be compared with those dominating it, hence the number of total comparisons as well as the time complexity could be decreased. To this end, ENS first sorts the population in an ascending order of the first objective value. In this way, solution p just needs to be compared with those in front of it in the sorted population, since those in the back of p have larger values on the first objective and can never dominate p. The procedure of ENS is described as follows, where P is the population to be sorted and $F = \{F_1, F_2, \ldots\}$ is the set of fronts.

Step 1: Sort P in an ascending order of the first objective value.
Step 2: For each solution p in the sorted population:

- (a) Assign p to F by comparing it with those having been assigned to F.

For assigning p to F, we must find its front F_i that satisfies

$$\begin{cases} \exists q \in F_{i-1}: & q \prec p,\ i > 1 \\ \nexists r \in F_i: & r \prec p \end{cases}$$

Therefore, the assignment of p is converted to a search problem. Two search strategies for this task, i.e. sequential search and binary search, will be given in the following section.

4.2.1.2. *ENS based on sequential search and binary search*

The ENS based on sequential search (ENS-SS) is to check from the first front F_1, until it finds a front F_i that satisfies the above condition, which works as follows:

Step 1: $i = 1$.
Step 2: If there is no solution in F_i dominates p, assign p to F_i and stop; otherwise $i = i + 1$ and go to Step 1.

The ENS based on binary search (ENS-BS) is to check from the middle front $F_{\lfloor K/2 \rfloor}$, where K denotes the number of fronts in the current F. ENS-BS works as follows:

Step 1: $lower = 0, upper = K, i = \lfloor (lower + upper + 1)/2 \rfloor$.
Step 2: If there is no solution in F_i dominates p, go to Step 3; otherwise, go to Step 4.
Step 3: If $i = lower{+}1$, assign p to F_i and stop; otherwise, $upper = i$, $i = \lfloor (lower + upper + 1)/2 \rfloor$ and go to Step 2.
Step 4: $lower = i$. If $upper = lower + 1$ and $upper < K$, assign p to F_{upper} and stop; else if $lower = K$, assign p to F_{K+1} and stop; otherwise, $i = \lfloor (lower{+}upper{+}1)/2 \rfloor$ and go to Step 2.

4.2.1.3. *Time complexity of ENS*

Sorting the population according to the first objective value can be achieved by the Heapsort, which holds a time complexity of $O(N \log N)$. In the worst case that all the solutions in the population are non-dominated mutually, each pair of solutions should be compared by ENS-SS and ENS-BS, so the number of comparisons is

$$S = \frac{N(N-1)}{2} = O(N^2)$$

and the worst case time complexity of ENS-SS and ENS-BS is

$$T_{\text{worst}} = O(N \log N) + MS = O(MN^2)$$

where M and N denote the number of objectives and the population size, respectively. Therefore, ENS has the same time complexity to fast non-dominated sort in the worst case.

For ENS-SS, it can obtain the best-case time complexity when there are $\sqrt{N}$ fronts with each containing $\sqrt{N}$ solutions. In this case, for each solution belonging to front F_i, it should be compared to at least one solution in each front F_j $(j < i)$, hence the number of comparisons is $\sqrt{N}(i-1)$ for all the solutions in F_i; on the other hand, the solutions in F_i should be compared mutually, which leads to $\sqrt{N}(\sqrt{N}-1)/2$ comparisons. To summarize, the best-case time complexity of ENS-SS is

$$\begin{aligned} T_{\text{best}}^{SS} &= O(N \log N) + M \cdot \sum_{i=1}^{\sqrt{N}} \left[\sqrt{N}(i-1) + \frac{\sqrt{N}(\sqrt{N}-1)}{2} \right] \\ &= O(N \log N) + M \cdot \left[\frac{N(\sqrt{N}-1)}{2} + \frac{N(\sqrt{N}-1)}{2} \right] \\ &= O(N \log N) + O(MN\sqrt{N}) \\ &= O(MN\sqrt{N}) \end{aligned}$$

For ENS-BS, it can obtain the best-case time complexity when there are N fronts with each containing one solution. In this case, $\lceil \log i \rceil$ fronts should be checked for assigning the solution belonging to front F_i. So the best-case time complexity of ENS-BS is

$$\begin{aligned} T_{\text{best}}^{BS} &= O(N \log N) + M \cdot \sum_{i=1}^{N} \lceil \log i \rceil \\ &= O(N \log N) + M \cdot \lceil \log(N!) \rceil \\ &= O(N \log N) + O(MN \log N) \\ &= O(MN \log N) \end{aligned}$$

As a consequence, the best-case time complexity of ENS is much smaller than $O(MN^2)$, hence it can work more efficiently than the fast non-dominated sort in most cases.

4.2.1.4. *Experimental results of ENS*

To verify the efficiency of ENS, it is compared to the fast non-dominated sort, arena's principle [16] and deductive sort [1] by being embedded in NSGA-II to solve DTLZ1 [5]. The population size of NSGA-II is set to 200 and 800, and the number of objectives of DTLZ1 is set to 2, 5 and 10. Each experiment is performed for 10 times, and the mean value of the runtime (in second) is recorded.

The experimental results are listed in Table 4.1. It can be seen that the runtime of ENS is much smaller than that of other non-dominated sorting approaches. For example, the runtime of ENS-BS

Table 4.1. Runtime of ENS, fast non-dominated, arena's principle and deductive sort in NSGA-II for solving DTLZ1.

Population size	Approach	Runtime (s) 2-objective	5-objective	10-objective
200	ENS-BS	**6.28**	28.94	45.36
	ENS-SS	6.36	**24.49**	**42.35**
	Deductive sort	12.35	32.76	52.88
	Arena's principle	17.16	47.39	70.03
	Fast non-dominated sort	102.92	110.81	124.15
800	ENS-BS	**34.06**	258.20	627.21
	ENS-SS	43.48	**192.21**	**499.30**
	Deductive sort	127.91	234.17	649.16
	Arena's principle	186.09	337.47	797.13
	Fast non-dominated sort	1674.6	1843.5	1858.2

is approximately 1/50 of the runtime of the fast non-dominated sort when the population size is 800 and the number of objectives is 2. In addition, the ENS-BS performs the best on 2-objective problem, and the ENS-SS performs the best on 5- and 10-objective problems.

4.2.2. *Tree-Based Efficient Non-dominated Sort (T-ENS)*

4.2.2.1. *Procedure of T-ENS*

According to the definition of Pareto dominance, the probability that an arbitrary solution dominates another one is $(1/2)^M$. So most solutions in a population are non-dominated mutually for an MOP where M is large. In this case, most non-dominated sorting approaches including ENS approximate to the same worse-case time complexity $O(MN^2)$. This phenomenon can be easily seen from Table 4.1, where the superiority of ENS over deductive sort decreases significantly with the increment of the number of objectives.

To address this issue, the tree-based efficient non-dominated sort (T-ENS) [22] employs a tree structure to store the assigned solutions, which can greatly decrease the runtime of non-dominated sorting on MaOPs. The trees in T-ENS have the following properties:

(1) They are all $(M-1)$-d trees, where M is the number of objectives.
(2) Each tree indicates a front, and each node in the tree indicates a solution having been assigned to the front.
(3) For each node r and its kth child q,

$$\begin{cases} f_i(r) \leq f_i(q), & \forall i = 1, \ldots, k \\ f_{k+1}(r) > f_{k+1}(q) \end{cases}$$

(4) Insertion of new nodes does not change the locations of existing nodes in the tree.

The third property is crucial to the tree, which makes it unnecessary to traverse all the existing nodes when inserting a new node. In other words, each solution needs not be compared with all the solutions in the same front, hence the number of comparisons of non-dominated sorting can be highly decreased. The fourth property guarantees that the insertion of new nodes holds a low time complexity, which can be achieved by the first operation of ENS, i.e. sorting the population in an ascending order of the first objective value.

When detecting whether a front has solution dominating p or not, ENS-SS and ENS-BS compare p with each solution in sequence, while in T-ENS, p only needs to be compared with several solutions in the tree. The traversal of the tree follows the following three rules:

(1) If the tree is empty, stop.
(2) Otherwise traverse the tree by preorder traversal, until finding a node dominating p.
(3) When the node r has been traversed, it only needs to traverse the first k subtrees of r, where k satisfies that

$$\begin{cases} f_i(r) \leq f_i(p), & \forall i = 1, \ldots, k-1 \\ f_k(r) > f_k(p) \end{cases}$$

The third rule provides a pruning strategy that makes p to be compared only with a part of the solutions in the tree, where this rule is established by the fact that there is no solution in the ith subtree of r dominating p when $i > k$. This fact can be easily demonstrated: According to the third property of the tree, $f_k(r) \leq f_k(q)$ holds for every node q in the ith subtree of r and $i > k$, so $f_k(p) < f_k(r) \leq f_k(q)$ and q can never dominate p.

After traversing a tree, if there is a solution in the tree dominating p, T-ENS will continue traversing the next tree for p; otherwise, it will insert p to the tree according to the following rules:

(1) If the tree is empty, insert p to the root of the tree.
(2) Otherwise insert p to the kth subtree of the root by the same strategy, where k satisfies that

$$\begin{cases} f_i(r) \leq f_i(p), & \forall i = 1, \ldots, k-1 \\ f_k(r) > f_k(p) \end{cases}$$

To summarize, T-ENS holds the same framework to ENS-SS, where the difference is that ENS-SS stores the solutions of each front in an array and compares p with each solution in the sequence, while T-ENS stores the solutions of each front in a tree and compares p with each solution by traversing the tree.

4.2.2.2. *Time complexity of T-ENS*

The sorting of the population according to the first objective value holds a time complexity of $O(N \log N)$. For the worse case of T-ENS, there are N fronts with each containing one solution. In this case, each pair of solutions needs to be compared and the tree structure is totally useless. So the worse-case time complexity of T-ENS is

$$\begin{aligned} T_{\text{worst}} &= O(N \log N) + M \cdot \frac{N(N-1)}{2} \\ &= O(N \log N) + O(MN^2) \\ &= O(MN^2) \end{aligned}$$

For the best case of T-ENS, there is only one front and the corresponding tree is full. Therefore, the depth K of the tree containing N nodes satisfies

$$\sum_{i=1}^{K} (M-1)^{i-1} = N$$

and

$$K = \lceil \log_{M-1}[N(M-2)+1] \rceil$$

For each solution in the ith layer of the tree, it should be compared with at least all its ancestors, which results in $i-1$ comparisons. So the number of total comparisons is

$$S = \sum_{i=1}^{K} (i-1)(M-1)^{i-1}$$

Then, we have

$$
\begin{aligned}
(M-1)S - S &= \sum_{i=1}^{K}(i-1)(M-1)^i - \sum_{i=1}^{K}(i-1)(M-1)^{i-1} \\
&= (K-1)(M-1)^K - \frac{(M-1)^K - (M-1)}{M-2}
\end{aligned}
$$

therefore

$$S = O\left(\frac{KM^K}{M-2}\right) = O\left(N\frac{\log N}{\log M}\right)$$

and the best-case time complexity of T-ENS is

$$T_{\text{best}} = O(N\log N) + MS = O\left(MN\frac{\log N}{\log M}\right)$$

As a result, the worse-case time complexity of T-ENS is the same as that of ENS-SS and ENS-BS, while the best-case time complexity of T-ENS is significantly smaller than the latter. More importantly, for MaOPs where most solutions are non-dominated mutually, ENS-SS and ENS-BS will hold a time complexity approximating the worst case, whereas T-ENS will hold a time complexity approximating the best case.

4.2.2.3. *Experimental results of T-ENS*

The efficiency of T-ENS is verified by being compared with the ENS-SS, fast non-dominated sort, corner sort [17] and deductive sort. The five non-dominated sorting approaches are embedded in KnEA [24] to solve DTLZ2 [5]. The population size of KnEA is set to 100 and 500, and the number of objectives of DTLZ2 is set to 3, 8 and 15.

It can be seen from the results shown in Table 4.2 that T-ENS consumes the least runtime on all the many-objective test instances. Moreover, the runtime of other non-dominated sorting approaches increases obviously when the number of objectives varies from 3 to 15, whereas the runtime of T-ENS fluctuates slightly with the increment of the number of objectives.

Table 4.2. Runtime of T-ENS, ENS-SS, fast non-dominated, corner sort and deductive sort in KnEA for solving DTLZ2.

Population size	Approach	Runtime (s)		
		3-objective	8-objective	15-objective
100	T-ENS	**4.93**	**4.44**	**5.71**
	ENS-SS	5.87	7.78	9.74
	Corner sort	9.44	11.38	13.66
	Deductive sort	9.14	11.25	13.58
	Fast non-dominated sort	21.04	23.06	27.44
500	T-ENS	**109.37**	**62.45**	**98.52**
	ENS-SS	135.62	160.65	221.79
	Corner sort	209.77	221.35	298.50
	Deductive sort	214.18	231.83	313.06
	Fast non-dominated sort	510.53	563.40	637.65

4.2.3. *Approximate Efficient Non-dominated Sort (A-ENS)*

4.2.3.1. *Procedure of A-ENS*

Since ENS sorts the population in an ascending order of the first objective value, the solution p being assigned can never dominate a solution q having been assigned. So when comparing p and q, it only needs to check whether there exists an objective where the value of p is smaller than the value of q. If such an objective exists, p will be non-dominated with q; otherwise, p will be dominated by q. Motivated by this idea, the approximate efficient non-dominated sort (A-ENS) [25] uses an approximate method to further increase the efficiency of T-ENS on MaOPs.

Specifically, A-ENS considers that p is non-dominated with q if one of the following three conditions holds:

$$f_m(p) < f_m(q), \quad m = \text{argmin}_{i=2,\ldots,M} f_i(p)$$

$$f_n(p) < f_n(q), \quad n = \text{argmax}_{i=2,\ldots,M} f_i(q)$$

$$\sum_{i=2}^{M} f_i(p) < \sum_{i=2}^{M} f_i(q)$$

Otherwise p is dominated by q. As a result, A-ENS holds the same framework to ENS-SS, but it modifies the definition of Pareto dominance, where p and q are compared on at most three objectives: the objective where p has the best value, the objective where q has the worst value and the sum of the objective values. The main reason for selecting only the three objectives is that the probability that p is smaller than q is relatively high on these three objectives, and the time complexity of each Pareto comparison is reduced from $O(M)$ to $O(1)$.

It is worth noting that the above three conditions are sufficient but not necessary for p being non-dominated with q, hence there may exist the case that p is considered to be dominated by q but it is actually non-dominated with q. This is the reason why A-ENS is called an approximate non-dominated sorting approach. Encouragingly, only a few solutions will be sorted incorrectly by A-ENS in practice, while it can bring impressive improvement to the efficiency.

4.2.3.2. *Time complexity of T-ENS*

The sorting of the population according to the first objective value holds a time complexity of $O(N \log N)$. The detection of the best value and worst value of each solution and the summing of objective values hold a time complexity of $O(MN)$. The worse case and best case of A-ENS are the same as those of ENS-SS, except the time complexity of each Pareto comparison is reduced from $O(M)$ to $O(1)$. So the worst-case time complexity of A-ENS is

$$T_{\text{worst}} = O(N \log N) + O(MN) + O(N^2) = O(N^2)$$

and the best-case time complexity of A-ENS is

$$T_{\text{best}} = O(N \log N) + O(MN) + O(N\sqrt{N}) = O(N\sqrt{N})$$

where it assumes that the number of objectives M is significantly smaller than the population size N. As a consequence, the time complexity of A-ENS is obviously smaller than other non-dominated sorting algorithms, including ENS-SS, ENS-BS and T-ENS in terms of both worst case and best case.

4.2.3.3. *Experimental results of A-ENS*

The efficiency of A-ENS is investigated by being compared to the T-ENS, ENS-SS, fast non-dominated sort and corner sort. These non-dominated sorting approaches are embedded in NSGA-III [3] to solve WFG1–WFG9 [13]. The population size of NSGA-III is set to 132, 156 and 275 for 6-, 8- and 10-objective problems, respectively.

Table 4.3 gives the average runtime of the five non-dominated sorting approaches in NSGA-III for solving WFG1–WFG9. It can be concluded from the experimental results that A-ENS is the most efficient non-dominated approach for solving MaOPs. It should be noted that other non-dominated sorting approaches are accurate methods, while A-ENS is an approximate method that improves the efficiency at the expense of the sorting accuracy. To investigate the influence of the sorting error on the effectiveness of NSGA-III, the performance of NSGA-III embedded with that of A-ENS is compared with that of NSGA-III embedded with T-ENS.

Table 4.4 lists the hypervolume [18] values of the results obtained by the two MOEAs on WFG1–WFG9, averaging over 10 runs. Note that a larger hypervolume value indicates a better performance of the algorithm. It is obvious that A-ENS dramatically improves the performance of NSGA-III to some extent, even if it cannot correctly sort the population during the evolution. The reason is that A-ENS may identify a few non-dominated solutions as dominated solutions, where these solutions generally have worse convergence than the others, so

Table 4.3. Runtime of A-ENS, T-ENS, ENS-SS, fast non-dominated and corner sort in NSGA-III for solving WFG1–WFG9.

	Runtime (s)		
Approach	6-objective	8-objective	10-objective
A-ENS	**8.97**	**12.06**	**35.06**
T-ENS	11.28	17.05	54.17
ENS-SS	21.76	35.29	123.37
Corner sort	29.16	44.97	152.29
Fast non-dominated sort	71.25	109.80	368.82

Table 4.4. Hypervolume of NSGA-III embedded with A-ENS and NSGA-III embedded with T-ENS on WFG1–WFG9 with 6-, 8- and 10-objectives.

	6-objective		8-objective		10-objective	
Problem	A-ENS	T-ENS	A-ENS	T-ENS	A-ENS	T-ENS
WFG1	**7.636e−1**	1.778e−1	**7.446e−1**	1.201e−1	**6.490e−1**	3.375e−2
WFG2	9.394e−1	**9.737e−1**	**9.940e−1**	9.923e−1	**9.979e−1**	9.919e−1
WFG3	5.542e−1	**5.561e−1**	**5.569e−1**	5.233e−1	**5.309e−1**	5.222e−1
WFG4	**6.641e−1**	6.107e−1	**7.719e−1**	7.020e−1	**8.493e−1**	7.567e−1
WFG5	**6.550e−1**	6.264e−1	**7.559e−1**	7.306e−1	**8.059e−1**	7.448e−1
WFG6	**6.435e−1**	6.181e−1	**7.327e−1**	6.980e−1	**7.875e−1**	7.206e−1
WFG7	**7.131e−1**	6.567e−1	**8.146e−1**	7.556e−1	**8.931e−1**	8.010e−1
WFG8	**5.425e−1**	4.770e−1	**6.305e−1**	6.073e−1	**7.138e−1**	6.634e−1
WFG9	**6.095e−1**	4.857e−1	**7.474e−1**	6.922e−1	**8.082e−1**	7.243e−1

the elimination of them can enhance the quality of the non-dominated solution set.

4.2.4. *Conclusion*

Since the non-dominated sorting was first employed by NSGA in 1995 [15], there exist approximately 20 non-dominated sorting approaches that have been proposed to improve the efficiency of MOEAs, where ENS is one of the most efficient non-dominated frameworks so far. Specifically, ENS-SS and ENS-BS are recommended for solving MOPs with two or three objectives, T-ENS is recommended for solving MOPs with more than three objectives (i.e. MaOPs), and A-ENS is recommended for improving both the efficiency and effectiveness of solving MaOPs. Besides, readers are recommended to Ref. [22] for a detailed introduction to all the existing non-dominated sorting approaches.

4.3. Improving the Effectiveness of Multiobjective Evolutionary Algorithms

The non-dominated sorting approaches introduced above can improve the efficiency of MOEAs, but they are not able to substantially improve the effectiveness when tackling complex MOPs, since the sorting results of different non-dominated sorting approaches besides A-ENS are identical. By contrast, the improvement of effectiveness should be brought by MOEAs. In this section, two MOEAs with novel search strategies are introduced for solving complex MOPs with many objectives and a large number of decision variables, respectively.

4.3.1. *Solving many-objective optimization problems*

4.3.1.1. *Framework of KnEA*

As discussed above, most solutions in a population are non-dominated mutually for an MaOP, hence MOEAs cannot distinguish most solutions by non-dominated sorting when solving MaOPs. To address this issue, the knee point-driven evolutionary algorithm (KnEA) [24] adopts the concepts of knee points to prefer those with better convergence among all the non-dominated solutions. Intuitively, a knee point in the Pareto front refers to the solution with the maximum marginal rates of return, which means that a small improvement in one objective of such a solution is accompanied by a severe degradation in at least another [2]. Therefore, knee points are with good convergence and naturally preferred.

KnEA follows the basic framework of most non-dominated sorting based MOEAs, which works as follows:

Step 1: Specify the population size N and the knee point ratio T.
Step 2: Randomly initialize a population with N solutions.
Step 3: For each generation:

- (a) Mating selection: Select a number of parents from the current population.
- (b) Generate offspring and combine them with the population.
- (c) Do non-dominated sorting by T-ENS.
- (d) Find the knee points in each front.
- (e) Environmental selection: Select N solutions from the combined population for the next generation.

There are three operations crucial to KnEA, i.e. mating selection, knee point identification and environmental selection. Besides, the other operations like initializing population and generating offspring are the same to most MOEAs such as NSGA-II [4].

4.3.1.2. *Mating selection of KnEA*

Mating selection is to select a number of parents for generating offsprings. The mating selection of KnEA uses the binary tournament selection to select the parents from the current population, with respect to three criteria, i.e. dominance comparison, knee point criterion and weighted distance. To be specific, each time KnEA randomly picks up two solutions from the population, and chooses the one dominating the other one. If the two solutions are non-dominated with each other and only one of them is a knee point, then the knee point is chosen. While if both of them are knee points or neither of them is a knee point, KnEA calculates the weighted distances of these two solutions, and chooses the one having a larger weighted distance value. The weighted distance of a solution p measures the quality of the solution in terms of diversity, which is calculated by

$$
\begin{aligned}
DW(p) &= \sum_{i=1}^{k} w_{p_i} \mathrm{dis}_{pp_i} \\
w_{p_i} &= \frac{r_{p_i}}{\sum_{i=1}^{k} r_{p_i}} \\
r_{p_i} &= \frac{1}{|\mathrm{dis}_{pp_i} - \frac{1}{k}\sum_{i=1}^{k} \mathrm{dis}_{pp_i}|}
\end{aligned}
$$

where p_i represents the ith nearest neighbour of p in the population, w_{p_i} represents the weight of p_i, dis_{pp_i} represents the Euclidean distance between p and p_i, and r_{p_i} represents the rank of distance dis_{pp_i} among all the distance dis_{pp_j}, $1 \leq j \leq k$. In practice, the parameter k is suggested to be 3.

4.3.1.3. *Adaptive knee point identification*

The knee points are used as the criterion next to the dominance relation in both mating selection and environmental selection, hence they play a central role in KnEA. Briefly, KnEA uses an adaptive strategy to identify all the knee points in each front. First of all, the M extreme solutions in each front are found (i.e. those having the maximum value on each of the M objectives), and used to construct a hyperplane L. Then, the Euclidean distance of each solution in the front to L is calculated; note that the MOP to be solved is assumed to be a minimization problem, hence the solutions lying on the same side of L to the origin are preferred. As a result, the distance between solution p and L can be calculated as

$$\text{dis}_{pL} = -\frac{f(p)^T X^{-1} e - 1}{\|X^{-1} e\|}$$

where X is a matrix with each row denoting one extreme solution and each column denoting one objective value, and e is a column vector whose elements are all set to one.

After obtaining the distance of each solution to the hyperplane, KnEA selects the knee points among the solutions in the front one by one. It first selects the one having the maximum distance to L as a knee point, and eliminates all the other solutions in its neighbourhood. Then it selects the one having the maximum distance to L among the remaining solutions, and eliminates the solutions in its neighbourhood as well. This procedure is repeated until all the solutions have been selected or eliminated.

The neighbourhood of each solution is defined by an M-dimensional hypercube with the solution being the centre. KnEA uses an adaptive method to update the neighbourhood size, which can make the ratio of the selected knee points approximately equal to the predefined value T. To be specific, the length of the neighbourhood on the ith dimension is calculated by

$$R_i = (\text{fmax}_i - \text{fmin}_i) \cdot r$$

where fmax_i and fmin_i represent the maximum and minimum values of the population on the ith objective, respectively, and the adaptive parameter r is calculated by

$$r = r \cdot e^{-\frac{1-t/T}{M}}$$

where t represents the ratio of selected knee points in the previous generation. Besides, r and t are initialized to 1 and 0, respectively. It can be seen from the above formula that if the ratio of selected knee points in the previous generation is smaller than T, the neighbourhood size will be shrunk and more solutions will be selected as knee points in the current generation; by contrast, if the ratio of selected knee points in the previous generation is larger than T, the neighbourhood size will be expanded and fewer solutions will be selected as knee points in the current generation. As a consequence, the neighbourhood size is adapted during the evolution to make the ratio of selected knee points approximatly equal to T.

4.3.1.4. *Environmental selection of KnEA*

Environmental selection is to select fitter solutions for the next generation. The environmental selection of KnEA begins with selecting all the solutions in the first k fronts and the knee points in the $(k + 1)$th front, where k denotes the maximum value satisfying $|\bigcup_{i=1}^{k} F_i| < N$. If the number of selected solutions is larger than N, those with smaller distances to the hyperplane L in the $(k+1)$th front will be eliminated; by contrast, if the number of selected solutions is smaller than N, KnEA will select the remaining solutions in $(k+1)$th front that have larger distances to the hyperplane L. Finally, the number of selected solutions is equal to N, and they survive for the next generation.

4.3.1.5. *Experimental results of KnEA*

To verify the effectiveness of KnEA in solving MaOPs, it is compared to MOEA/D [20], GrEA [19] and NSGA-III [3] on WFG1–WFG9 with 6- and 10-objectives. Table 4.5 gives the hypervolume values of

Table 4.5. Hypervolume of MOEA/D, GrEA, NSGA-III and KnEA on WFG1–WFG9 with 6- and 10-objectives.

Problem	Obj.	MOEA/D	GrEA	NSGA-III	KnEA
WFG1	6	9.406e−1	9.754e−1	7.302e−1	**9.895e−1**
	10	8.376e−1	9.873e−1	4.914e−1	**9.944e−1**
WFG2	6	7.678e−1	9.315e−1	9.568e−1	**9.888e−1**
	10	9.304e−1	9.781e−1	**9.966e−1**	9.932e−1
WFG3	6	5.727e−1	**5.891e−1**	5.562e−1	5.485e−1
	10	**6.018e−1**	6.001e−1	6.005e−1	5.576e−1
WFG4	6	2.519e−1	6.238e−1	5.853e−1	**6.257e−1**
	10	4.033e−1	8.174e−1	7.852e−1	**8.377e−1**
WFG5	6	2.445e−1	**6.092e−1**	6.014e−1	6.068e−1
	10	3.197e−1	7.863e−1	7.837e−1	**8.122e−1**
WFG6	6	3.200e−1	**5.972e−1**	5.934e−1	5.890e−1
	10	3.601e−1	7.685e−1	7.714e−1	**7.883e−1**
WFG7	6	3.626e−1	6.852e−1	6.496e−1	**6.881e−1**
	10	3.705e−1	8.729e−1	8.513e−1	**8.948e−1**
WFG8	6	1.902e−1	**4.684e−1**	4.343e−1	4.353e−1
	10	3.128e−1	7.117e−1	6.625e−1	**7.150e−1**
WFG9	6	1.639e−1	5.768e−1	4.823e−1	**5.982e−1**
	10	3.088e−1	7.994e−1	7.552e−1	**8.077e−1**

the results obtained by the four MOEAs, averaging over 20 runs. It is clear that KnEA wins on 12 out of 18 test instances, while MOEA/D, GrEA and NSGA-III obtain only 1, 4 and 1 best results, respectively. In particular, KnEA outperforms all the peer algorithms on all the test instances with 10-objectives besides WFG2 and WFG3, so it can be concluded that KnEA is quite effective in solving MOPs with many objectives.

4.3.2. *Solving large-scale multiobjective optimization problems*

4.3.2.1. *Framework of LMEA*

In spite of the various approaches that are focused on the scalability of MOEAs to the number of objectives, little work has been reported to consider the large number of decision variables in multiobjective optimization. Generally, MOEAs with any novel selection strategies cannot work well on large-scale MOPs, which is mainly due to the fact that the decision space is too large for MOEAs to find solutions approximating the Pareto front. Therefore, it is desirable to adopt the divide-and-conquer methodology to divide the decision space and optimize each part separately.

The large-scale many-objective evolutionary algorithm (LMEA) [23] uses a decision variable clustering method to divide the decision variables into convergence-related and diversity-related ones, where each type of decision variables is optimized by a different strategy. The procedure of LMEA is listed as follows:

Step 1: Specify the population size N and other parameters.
Step 2: Randomly initialize a population with N solutions.
Step 3: Do variable clustering to divide the decision variables into convergence-related and diversity-related ones.
Step 4: Do interaction analysis to divide the convergence-related variables into independent subgroups.
Step 5: For each generation:

- (a) Optimize the convergence-related variables.
- (b) Optimize the diversity-related variables.

The details of variable clustering, interaction analysis and the two optimization strategies are elaborated in the following sections.

4.3.2.2. *Decision variable clustering*

For each decision variable, LMEA randomly picks up a solution from the population and performs a number of perturbations on the variable of the picked solution. Then a number of solutions having the same values on all the decision variables besides the perturbed one are obtained, and a line L is generated to fit the normalized objective values of these solutions as close as possible. Afterwards, the angle between the fitted line L and the normal line of hyperplane $f_1 + \cdots + f_M = 1$ is obtained to be the feature of the decision variable. Each decision variable finally has several features after picking up multiple solutions, and the k-means clustering method is adopted to divide the decision variables into two clusters based on the features of each variable.

To summarize, the decision variables are divided based on the directions of their corresponding lines. The fitted lines represent the direction of convergence, more specifically, a smaller angle between the fitted line and the normal line of the hyperplane indicates a higher probability that the corresponding decision variable is convergence-related, and vice versa.

4.3.2.3. *Interaction analysis of convergence-related variables*

After dividing the decision variables into two types, all the convergence-related variables are further divided into several subgroups based on the interactions between them, which is defined as follows:

Definition 4.7. Given a solution x and four values a_1, a_2, b_1 and b_2, decision variable i is said to interact with decision variable j if and only if there exists at least one objective k satisfying the following condition:

$$\begin{cases} f_k(x)|_{x_i=a_1,x_j=b_1} < f_k(x)|_{x_i=a_2,x_j=b_1} \\ f_k(x)|_{x_i=a_1,x_j=b_2} > f_k(x)|_{x_i=a_2,x_j=b_2} \end{cases}.$$

LMEA analyses the interaction between each pair of convergence-related variables by the above definition, where the solution x is randomly picked up from the population and a_1, a_2, b_1 and b_2 are set to random values inside the decision space. Afterwards, all the convergence-related variables are divided into several subgroups, where a variable is interacted with at least one variable inside the same subgroup but not interacted with those in any other subgroups. The variables in each subgroup cannot be optimized separately due to the interactions between them, whereas each subgroup of variables can be optimized one by one to decrease the search space.

4.3.2.4. *Two optimization strategies of LMEA*

Once the interaction analysis of convergence-related variables is done, LMEA alternately optimizes the convergence-related variables and the diversity-related variables by different strategies. For the convergence-related variables, LMEA optimizes each subgroup of variables one by one using a convergence optimization strategy. To be specific, a number of parents are selected from the current population using the binary tournament, and offspring solutions are generated by replacing the values of the convergence-related variables in the current subgroup with those generated by genetic operators, while leaving the remaining variables unchanged. For each offspring, it can replace its parent if and only if it has a smaller non-dominated front number, or it has the same non-dominated front number but a smaller Euclidean distance to the origin.

As for the optimization of diversity-related variables, the parents are randomly picked from the current population, and offspring solutions are generated by replacing the values of all the diversity-related variables with those generated by genetic operators. Then the population is combined with the offspring and sorted by T-ENS. LMEA selects all the solutions in the first k fronts with k denoting the maximum value satisfying $|\bigcup_{i=1}^{k} F_i| < N$, and selects the solutions of the best diversity one by one from the $(k+1)$th front until the number of selected solutions reaches N, where the diversity of each solution is measured by the angles between the solution and those having been selected in objective space.

4.3.2.5. *Experimental results of LMEA*

The effectiveness of LMEA in solving large-scale MOPs is verified by empirically comparing it with MOEA/D [20], NSGA-III [3], KnEA, and MOEA/DVA [8]. The DTLZ1–DTLZ7 [5] are used as the test problem in this experiment; the number of objectives of the problems is set to 5, and the number of decision variables is set to 500 and 1000. Table 4.6 gives the statistical results in terms of IGD [26], where a smaller IGD value indicates a better performance of the algorithm. As can be seen from the table, LMEA performs the best on 10 out of 14 test instances, hence it can be verified that LMEA is a promising algorithm for solving large-scale MOPs.

Table 4.6. IGD of MOEA/D, NSGA-III, KnEA, MOEA/DVA and LMEA on DTLZ1–DTLZ7 with 500 and 1000 decision variables.

Problem	Dec.	MOEA/D	NSGA-III	KnEA	MOEA/DVA	LMEA
DTLZ1	500	1.465e−1	5.260e+1	7.644e+1	6.328e−2	**6.112e−2**
	1000	1.683e−1	1.167e+2	1.340e+2	6.344e−2	**6.042e−2**
DTLZ2	500	3.201e−1	1.949e−1	2.363e−1	1.949e−1	**1.883e−1**
	1000	3.201e−1	1.949e−1	2.284e−1	1.949e−1	**1.882e−1**
DTLZ3	500	4.178e−1	2.942e+2	6.123e−1	1.954e−1	**1.904e−1**
	1000	4.468e−1	6.823e+2	8.248e−1	1.956e−1	**1.881e−1**
DTLZ4	500	5.273e−1	**1.950e−1**	2.157e−1	3.442e−1	2.726e−1
	1000	4.385e−1	2.283e−1	**2.138e−1**	3.442e−1	2.707e−1
DTLZ5	500	4.516e−2	1.941e−1	3.174e−1	2.047e−1	**4.086e−3**
	1000	4.516e−2	2.061e−1	3.891e−1	2.046e−1	**4.073e−3**
DTLZ6	500	1.301e+0	4.993e−1	7.275e−1	1.824e−1	**4.513e−3**
	1000	2.714e+0	6.577e−1	1.509e+0	1.824e−1	**3.975e−3**
DTLZ7	500	5.154e−1	2.393e+0	**2.319e−1**	5.204e−1	3.203e−1
	1000	5.212e−1	2.663e+0	**2.341e−1**	5.204e−1	3.105e−1

4.3.3. *Conclusion*

This section introduced several algorithms for handling different types of complex MOPs. For the MOPs with many objectives, the non-dominated sorting is ineffective to distinguish most solutions in the population, so KnEA favours the knee points among the non-dominated solutions to increase the selection pressure. For the MOPs with a large number of decision variables, the decision space is too large for algorithms to find well-converged solutions, hence LMEA adopts the divide-and-conquer methodology to divide the decision variables into different groups and optimizes them separately.

Bibliography

[1] Clymont, K. M. and Keedwell, E. Deductive sort and climbing sort: New methods for non-dominated sorting, *Evol. Comput.* **20**(1), 1–26, 2012.

[2] Das, I. On characterizing the "knee" of the Pareto curve based on normal-boundary intersection, *Struct. Optim.* **18**(2–3), 107–115, 1999.

[3] Deb, K. and Jain, H. An evolutionary many-objective optimization algorithm using reference-point based non-dominated sorting approach, part I: Solving problems with box constraints, *IEEE Trans. Evol. Comput.* **18**(4), 577–601, 2014.

[4] Deb, K., Pratap, A., Agarwal, S. and Meyarivan, T. A fast and elitist multi-objective genetic algorithm: NSGA-II, *IEEE Trans. Evol. Comput.* **6**(2), 182–197, 2002.

[5] Deb, K., Thiele, L., Laumanns, M. and Zitzler, E. Scalable test problems for evolutionary multiobjective optimization. In *Evolutionary Multiobjective Optimization.* Theoretical Advances and Applications, pp. 105–145, Springer, London, 2005.

[6] Handl, J. and Knowles, J. An evolutionary approach to multiobjective clustering. *IEEE Trans. Evol. Comput.* **11**(1), 56–76, 2007.

[7] Lazzerini, B., Marcelloni, F. and Vecchio, M. A multi-objective evolutionary approach to image quality/compression trade-off in jpeg baseline algorithm. *Appl. Soft Comput.* **10**(2), 548–561, 2010.

[8] Ma, X., Liu, F., Qi, Y., Wang, X., Li, L., Jiao, L., Yin, M. and Gong, M. A multiobjective evolutionary algorithm based on decision variable analyses for multiobjective optimization problems with large-scale variables. *IEEE Trans. Evol. Comput.* **20**(2), 275–298, 2016.

[9] Mukhopadhyay, A., Maulik, U., Bandyopadhyay, S. and Coello, C. A. C. A survey of multiobjective evolutionary algorithms for data mining: Part I, *IEEE Trans. Evol. Comput.* **18**(1), 4–19, 2014.

[10] Panduro, M. A., Brizuela, C. A., Covarrubias, D. and Lopez, C. A trade-off curve computation for linear antenna arrays using an evolutionary multi-objective approach. *Soft Comput.* **10**(2), 125–131, 2006.

[11] Pettersson, F., Chakraborti, N. and Saxén, H. A genetic algorithms based multi-objective neural net applied to noisy blast furnace data. *Appl. Soft Comput.* **7**(1), 387–397, 2007.

[12] Ponsich, A., Jaimes, A. L. and Coello, C. A. C. A survey on multiobjective evolutionary algorithms for the solution of the portfolio optimization problem and other finance and economics applications. *IEEE Trans. Evol. Comput.* **17**(3), 321–344, 2013.

[13] Huband, L. B.S., Hingston, P. and While, L. A review of multiobjective test problems and a scalable test problem toolkit. *IEEE Trans. Evol. Comput.* **10**(5), 477–506, 2006.

[14] Shin, S.-Y., Lee, I.-H., Kim, D. and Zhang, B.-T. Multiobjective evolutionary optimization of DNA sequences for reliable DNA computing. *IEEE Trans. Evol. Comput.* **9**(2), 143–158, 2005.

[15] Srinivas, N. and Deb, K. Multiobjective optimization using nondominated sorting in genetic algorithms. *Evol. Comput.* **2**(3), 221–248, 1995.

[16] Tang, S., Cai, Z. and Zheng, J. A fast method of constructing the non-dominated set: Arena's principle. In *Proceedings of the Fourth International Conference on Natural Computation*, pp. 391–395, 2008.

[17] Wang, H. and Yao, X. Corner sort for Pareto-based many-objective optimization. *IEEE Trans. Cybern.* **44**(1), 92–102, 2014.

[18] While, L., Hingston, P., Barone, L. and Huband, S. A faster algorithm for calculating hypervolume. *IEEE Trans. Evol. Comput.* **10**(1), 29–38, 2006.

[19] Yang, S., Li, M., Liu, X. and Zheng, J. A grid-based evolutionary algorithm for many-objective optimization, *IEEE Trans. Evol. Comp.* **17**(5), 721–736, 2013.

[20] Zhang, Q. and Li, H. MOEA/D: A multi-objective evolutionary algorithm based on decomposition, *IEEE Trans. Evol. Comput.* **11**(6), 712–731, 2007.

[21] Zhang, X. Tian, Y. Cheng, R. and Jin, Y. An efficient approach to non-dominated sorting for evolutionary multi-objective optimization. *IEEE Trans. Evol. Comput.* **19**(2), 201–213, 2015.

[22] Zhang, X., Tian, Y., Cheng, R. and Jin, Y. Empirical analysis of a tree-based efficient non-dominated sorting approach for many-objective optimization. In *Proceedings of the 2016 IEEE Symposium Series on Computational Intelligence*, 2016.

[23] Zhang, X., Tian, Y., Cheng, R. and Jin, Y. A decision variable clustering-based evolutionary algorithm for large-scale many-objective optimization. *IEEE Trans. Evol. Comput.* **22**(1), 97–112, 2018.

[24] Zhang, X., Tian, Y. and Jin, Y. A knee point driven evolutionary algorithm for many-objective optimization. *IEEE Trans. Evol. Comput.* **19**(6), 761–776, 2015.

[25] Zhang, X., Tian, Y. and Jin, Y. Approximate non-dominated sorting for evolutionary many-objective optimization. *Inf. Sci.* **369**, 14–33, 2016.

[26] Zhou, A., Jin, Y., Zhang, Q., Sendhoff, B. and Tsang, E. Combining model-based and genetics-based offspring generation for multi-objective optimization using a convergence criterion. In *Proceedings of the 2006 IEEE Congress on Evolutionary Computation*, pp. 892–899, 2006.

[27] Zhou, A., Qu, B.-Y., Li, H., Zhao, S.-Z., Suganthan, P. N. and Zhang, Q. Multiobjective evolutionary algorithms: A survey of the state of the art. *Swarm Evol. Comput.* **1**(1), 32–49, 2011.

Chapter 5

A Bio-inspired Clustering Model for Anomaly Detection in the Mining Industry

Raymond Chiong[*,§], Zhongyi Hu[†,¶], Zongwen Fan[*,||]
Yuqing Lin[*], Stefan Chalup[*] and Antoine Desmet[‡]

[*]*School of Electrical Engineering and Computing*
The University of Newcastle, Callaghan
NSW 2308, Australia

[†]*School of Information Management*
Wuhan University, Wuhan 430072, China

[‡]*Komatsu Mining Corp. Group (formerly*
Joy Global Inc.), 95 Mustang Drive, Rutherford
NSW 2320, Australia

[§]*raymond.chiong@newcastle.edu.au*
[¶]*Zhongyi.hu@whu.edu.cn*
[||]*Zongwen.Fan@uon.edu.au*

Being able to detect anomalies for predicting machine breakdown is of critical importance in the mining industry. These anomalies are usually found in unlabelled sensor data, and therefore unsupervised models represent the preferred choice for the task. In this chapter, we propose the use of a bio-inspired clustering model based on the self-organizing map (SOM) to detect anomalies in real-world data provided by Joy Global, a manufacturer of high-productivity mining solutions. The proposed SOM is compared to two other well-known clustering models, namely k-means and fuzzy c-means. Simulation experiments using grease cycle data from the manufacturer show that the SOM is able to detect a more reasonable number of anomalies than k-means and fuzzy c-means. Based on real scenarios given by Joy Global, we devise a simple way to prevent machine failures by triggering alarms through the anomalies detected, and the SOM is again shown to be more capable of identifying incidents of potential machine breakdown compared to the other two clustering models.

5.1. Introduction

The mining industry is highly reliant on sensors to monitor their mining equipment. With the help of sensors, the behaviour and operational profile of mining machines are monitored to avoid undesired failures and prevent costly interruption. When the volume of sensor data is large, however, manually checking for anomalies in the data can be extremely challenging. Being able to automatically detect anomalies in this context is thus crucial for monitoring the condition of machines [12, 20]. With accurate detection of anomalies in the machines, decision makers can be provided with ample warning and given enough time to adequately act on, prevent or correct situations associated with the equipment or system.

Generally speaking, anomaly detection aims to find patterns that exhibit abnormal or unexpected behaviour (i.e. anomalies or outliers) in a dataset. If the data is sufficiently balanced between labelled normal and abnormal samples, supervised learning (with weight balancing) can be applied to train classifiers such as the support vector machine (SVM), multilayer perceptron and ensemble machines [13, 35]. In the mining industry, however, normal and abnormal samples are usually unlabelled. Furthermore, it is very expensive to label normal and abnormal samples in this case. There are also factors such as temperature and moisture that would cause anomalies in mining equipment. An anomaly can occur over days, minutes or even suddenly, making it difficult to detect [1, 4, 5]. Instead of trying to build supervised models to classify anomalies, here unsupervised models are our preferred choice.

Many unsupervised models for detecting anomalies exist in the relevant literature. These include clustering models, which cluster data into two separate groups, namely *normal* and *abnormal* groups [5, 35]. Support vector clustering was proposed by

Ben-Hur *et al.* [2]. It was later accelerated in several follow-up papers [18, 26, 27, 31] and was more recently applied to time series data using kernels based on dynamic time warping [3, 16]. Möller-Levet *et al.* [21] used a fuzzy c-means (FCM) clustering algorithm for short time series and unevenly distributed data points. A linear one-class SVM with deep learning was used by Erfani *et al.* [9] to deal with high-dimensional and large-scale anomaly detection. Girdhar *et al.* [11] applied Bayesian non-parametric scene modelling for anomaly detection in unstructured environments. Yassin *et al.* [32] investigated anomaly-based intrusion detection using k-means and naive Bayes algorithms. Samantaray [28] also studied anomaly intrusion detection but using a genetic-fuzzy rule mining algorithm. Model extraction and model checking were used by Ogawa *et al.* [23] to deal with experimental fault analysis. Anomaly detection on unlabelled data was studied by Eskin *et al.* [28] using a geometric framework. For more details about clustering algorithms for anomaly detection, see [1, 22, 33].

In this chapter, we present a bio-inspired clustering model based on the self-organizing map (SOM) to detect anomalies in mining equipment. The SOM is a bio-inspired artificial neural network (ANN) considering the notion of self-organization and spatial organization of information within the brain [17, 29, 30]. Unlike traditional ANNs, which originated from the functionality and interoperability of neurons, the SOM was inspired by the existence of "maps" that represent spatially organized responses within the brain. With the concept of "maps", two-dimensional maps representing a scaled version of n-dimensional data are used as input for the proposed model. It has the ability to project input into a low-dimensional space, so that even high-dimensional data can be visualized but still maintain the topographic relationships of data. Based on a competitive learning mechanism with self-organizing properties, the SOM has been widely applied in data mining scenarios [29].

As a case study, we use grease cycle data from Joy Global for our computational experiments. Joy Global is a world-renowned leader in manufacturing, delivering high-productivity mining solutions.

The JoySmart Remote Machine Health Monitoring Centre of Joy Global collects and processes sensor data from their customers' surface and underground mining equipment. The Centre's mission is to detect the onset of any faults that may occur as early as possible. The grease cycle data used for our computational experiments were collected by sensors to record the cycle time of grease pumps that push grease through a network of hoses to grease various parts of the machines. It is very important to detect anomalies in grease pumps based on their sensor data. In this study, the SOM is applied to cluster grease cycle data into two groups. Two well-known clustering algorithms, namely k-means and FCM, are also used for comparison purposes. Experimental results show that the SOM can detect a more reasonable number of anomalies in grease pumps compared to k-means and FCM, which may provide decision makers essential support in monitoring the health of their mining machines.

The remainder of this chapter is organized as follows. In Section 5.2, we describe the grease cycle data used and its source in detail. Then, clustering models including the SOM, k-means and FCM are presented in Section 5.3. After that, we discuss the experimental setup and results in Section 5.4, before conclusion is drawn in Section 5.5.

5.2. Data and Description

Among the tons of sensor data Joy Global has, this study will focus on a group of sensors related to grease pumps, which are an important part of their mining equipment. We first select relevant data from the sensor data provided, and then apply preprocessing techniques in order to remove noise within the data as well as to reduce the dimensionality. Due to the environments in which the sensors are running, the readings can be affected by numerous factors, such as humidity and pressure. In the following sections, we will explain the source of data and carry out some descriptive analysis.

5.2.1. *Data source*

Joy Global's JoySmart Remote Machine Health Monitoring Centre collects and processes a huge volume of sensor data from their customers' surface and underground mining equipment. Their mission is to detect the onset of any faults that may occur as early as possible. This will give customers sufficient time to plan for maintenance, order parts and prevent any major costly interruption to the mining process. The main challenge for Joy Global is to deal with the large amount of data efficiently. Each mining machine is able to transmit up to 700 sensor readings every second, and Joy Global is currently monitoring more than 30 machines. A computer-based solution is thus preferred, where computer programs can scan the input data streams from each of the sensors to detect any anomalies. Given the considerable scale of mining machinery, any maintenance necessary is not a trivial undertaking. Significant costs are associated with such maintenance. It is therefore vital that accurate fault predictions are undertaken, i.e. false alarms are not acceptable.

In this study, we focus on data related to the cycle time of grease pumps. A grease pump pushes grease through a network of hoses, in order to grease various parts of the machines. The cycle time measures how long the pump takes to reach a certain pressure inside these hoses. Once the pressure is reached, the cycle is considered complete, and the hoses are depressurized before they go into the next cycle.

There are two types of cycles, which are activated alternately (e.g. the lower type is first activated, then nothing, then upper, then nothing, and the cycle repeats). Both types of cycles are powered by the same pump, so any issue with the pump would affect them both. The pump aside, these cycles are independent, and any of them could develop faults that are only visible when they are activated.

5.2.2. *Descriptive analysis*

The grease cycle data used in this study was collected from February 10, 2015 to January 23, 2017. During a grease cycle, a grease pump draws grease out of a grease tank, and pumps it into a network of pipes. The pipes deliver fresh grease to moving parts of a machine. A grease cycle happens every 15–30 minutes. What is measured in a grease cycle is the time (in seconds) the pump takes to pump grease from 0 PSI (pounds per square inch) up to a set pressure of 2200 PSI (the pressure measured inside grease pipes).

Table 5.1 and Fig. 5.1 show examples of the grease cycle data. The duration of cycles exhibits some patterns when the time and type of cycles change. What we want to find are hidden patterns so that anomalies can be detected.

If there are no leaks, no obstructions, and the grease pump is not deteriorating, the cycles should take about the same time to complete. However, typical failures that often come up include:

(1) A leak in the network of grease pipes: It will be hard for the pump to reach 2200 PSI, if the grease being pumped is escaping out. The larger the leak, the longer it will take to

Table 5.1. First 20 samples of the grease cycle data.

Index	Time	Type of cycles	Duration of cycles
1	2015/2/10 6:23	Lower	21.1
2	2015/2/10 6:24	Upper	14.8
3	2015/2/10 6:40	Lower	21.6
4	2015/2/10 6:40	Upper	15.9
5	2015/2/10 6:55	Lower	22.2
6	2015/2/10 6:56	Upper	14.9
7	2015/2/10 7:01	Lower	21.2
8	2015/2/10 8:33	Lower	24
9	2015/2/10 8:42	Lower	23
10	2015/2/10 8:42	Upper	15.9
11	2015/2/10 8:58	Lower	22
12	2015/2/10 8:58	Upper	15.3
13	2015/2/10 9:13	Lower	20.2
14	2015/2/10 9:14	Upper	14.9
15	2015/2/10 9:32	Lower	21.6
16	2015/2/10 9:32	Upper	14.9
17	2015/2/10 9:47	Lower	20
18	2015/2/10 9:47	Upper	14.9
19	2015/2/10 9:59	Lower	20.1
20	2015/2/10 10:11	Lower	21.1

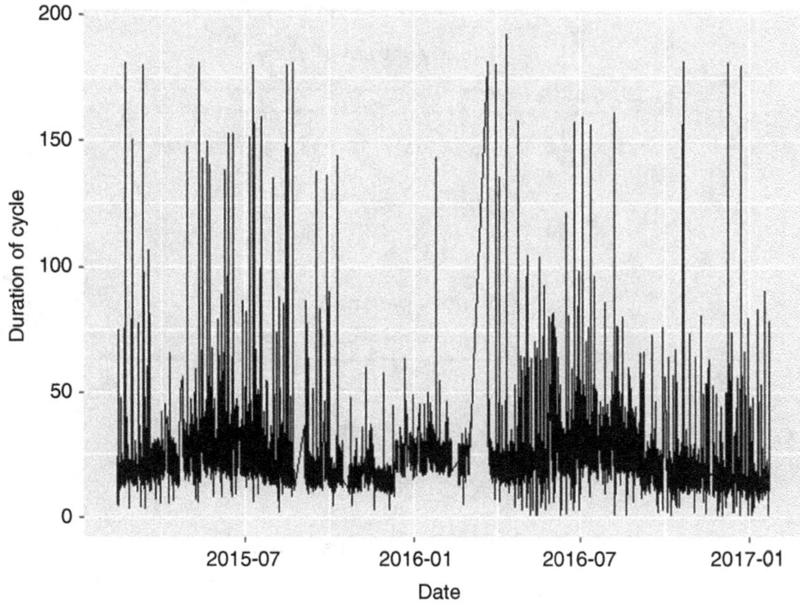

Fig. 5.1. A plot of the grease cycle data.

reach the set pressure. The grease cycle time will gradually increase.

(2) A failing grease pump: If the grease pump is getting worn out, it does not pump grease as efficiently. The cycle time will gradually take longer.

From Fig. 5.1, we can clearly see that there are many large values, indicating possible anomalies in the greasy pump system.

To analyze the pattern of grease cycle time over hours, days and months, the average values for each hour, day and month are calculated and shown in Fig. 5.2. One might observe an oscillation with hourly and monthly patterns. This is due to ambient temperature: hotter during the day and summer, colder during the night and winter. Hotter grease becomes more fluid, and thus easier to pump, which leads to quicker cycle times and vice versa during the night. However, these variations are much smaller than the ones that can be seen in Fig. 5.1 if there is a fault developing.

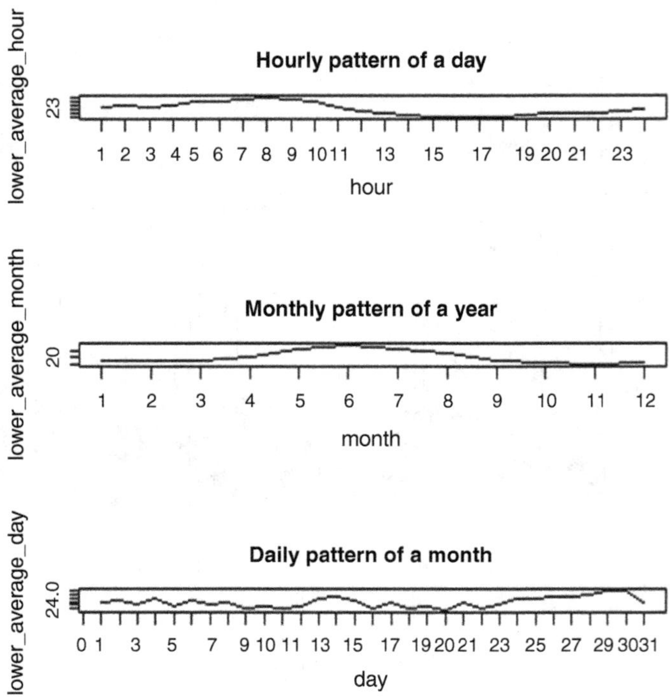

Fig. 5.2. Hourly, monthly and daily patterns of the grease cycle data.

5.3. Clustering Models

5.3.1. *Self-organizing map*

The self-organizing map (SOM) is a clustering algorithm based on an ANN [30]. It is divided into two layers: input and Kohonen layers. The SOM is an effective method to visualize high-dimensional data by reducing data dimensions to a map. It uses a competitive learning technique [19]: when input a sample, the winning unit is chosen from a weight vector that is closest to the current sample. In each iteration, neighbourhood relationships of the input data are preserved while adjusting the input variables. In other words, the neighbourhood function can preserve topological relationships of the input variables [10]. Basic procedures of the SOM are shown below:

Step 1: Start the iterations by initializing iteration t from 0 to *Iter*; randomly set the weight vector, $w(0)$; and initialize the learning rate, α_0.

Step 2: Input a sample from the dataset.

Step 3: Obtain the best match. Compute the distance between the current sample and neurons from the output layer. The neuron that is closest to the current sample is the best match.

Step 4: In each iteration, all parameters are adjusted. The following equations are used to update the parameter vector:

$$w_i(t) = w_i(t-1) + \alpha(t)\ \mathrm{neigh}_{y(\mathrm{win}),i}(t)(x - w_i(t)) \quad (5.1)$$

where $\mathrm{neigh}_{y(\mathrm{win}),i}(t)$ is the neighbourhood function. Here, we use the Gaussian function:

$$\mathrm{neigh}_{y(\mathrm{win}),i}(t) = \exp\left(-\frac{\left\|r_i - r_{y(\mathrm{win})}\right\|}{2\delta^2(t)}\right) \quad (5.2)$$

where r_i and $r_{y(\mathrm{win})}$ are the ith output coordinate and the best match coordinate, respectively. $\delta(t)$ decreases with t. When $\delta(t)$ is close to 0, $\mathrm{neigh}_{y(\mathrm{win}),i}(t)$ converges.

Step 5: Repeat steps 2–4 until all data points have been traversed.

Step 6: Repeat steps 2–5 until t reaches *Iter*.

Figure 5.3 shows a flowchart of the SOM algorithm used for anomaly detection in our study. Here, the number of neurons K is set to 2. However, this parameter can be changed using different values. This is a very important part of the training process, to train a good model and ensure good performance when the model is applied to testing data.

5.3.2. *K-means*

K-means [15] is a commonly used clustering algorithm. It divides the data into K clusters, to minimize the within-cluster sum of squares. Basic procedures of k-means are shown below:

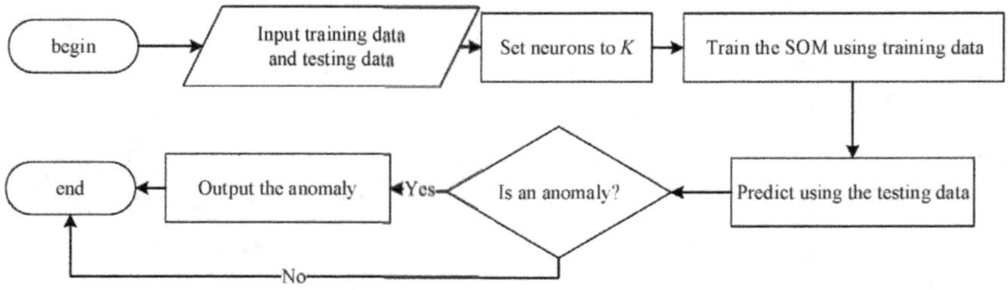

Fig. 5.3. A flowchart of the SOM algorithm used for anomaly detection in this study.

Step 1: Randomly choose Kcluster centres $u_1, u_2, \ldots, u_k$.

Step 2: For each data point, calculate the distances to all cluster centres, then assign this point to the closest cluster

$$c^{(i)} = \underset{j}{\operatorname{argmin}} \|x^{(i)} - u_j\| \tag{5.3}$$

where i is the index of each data point, and j is the index of cluster centres.

Step 3: For each cluster, recalculate the centre of this cluster

$$u^{(i)} = \frac{\sum_{i=1}^{m} 1\{c^{(i)} = j\}x^{(i)}}{\sum_{i=1}^{m} 1\{c^{(i)} = j\}} \tag{5.4}$$

where i is the index of each data point, and j is the index of cluster centres.

Step 4: Repeat steps 2–3 until the termination condition is met, i.e. the cluster centres are not changed or the maximum iteration is reached.

K-means is a very easy algorithm to understand and implement. However, it is sensitive to noise and the initial cluster centres chosen.

5.3.3. *Fuzzy c-means*

The fuzzy c-means (FCM) clustering algorithm [8, 24, 25, 34] can be used to identify inherent relationships in a large number of samples. Unlike k-means where each sample belongs to only one cluster, FCM assigns a sample to multiple clusters based on fuzzy memberships.

An optimized sample set and the distribution of samples can be visualized with the application of FCM. Like k-means, however, FCM is also affected by the initial cluster centres chosen.

Suppose that there is a set, $X = \{x_1, x_2, \ldots, x_N\}$, and each x_i has n attributes, represented as $x_i = \{x_{1i}, x_{2i}, \ldots, x_{ni}\}^T$. Dividing the data into c classes, the fuzzy c-partitions can be represented by a $c \times N$ Boolean matrix, $U = [u_{ij}]_{c\times N}$. The fuzzy partition space is shown as

$$M_{fc} = \left\{ U \subset R^{cN} \mid u_{ij} \in [0,1], \forall i, \forall j; \right.$$

$$\left. \sum_{j=1}^{c} u_{ij} = 1, \forall j; 0 < \sum_{i=1}^{N} u_{ij}, \forall j \right\} \tag{5.5}$$

The establishment of FCM is to get a $c \times N$ membership matrix $U = [u_{ij}]$ and c cluster centres:

$$J_m = \min \sum_{i=1}^{N} \sum_{j=1}^{c} \left[(u_{ij})^m d^2(x_i, \nu_j) \right] \tag{5.6}$$

Here, $\sum_{j=1}^{c} u_{ij} = 1, \forall i; 0 < \sum_{i=1}^{N} u_{ij} \leq N, 0 \leq u_{ij}$, $i = 1, 2, \ldots, N, j = 1, 2, \ldots, c$; u_{ij} represents the membership degree of sample data x_i in class j, and $m \in [1, +\infty]$ is the coefficient of fuzzy weighting; $d(x_i, v_j)$ corresponds to the distance between sample data x_i and cluster centre v_j; m in the formula, known as the weighted index, is set to 2 in this study.

J_m in Eq. (5.6) is the objective function of FCM. Cluster centres are updated through an iterative process. Basic procedures of FCM are shown below:

Step 1: Initialization. Set the number of cluster centres, c, and termination condition threshold value ε. Set the counter, b, to 0, and initialize membership matrix $U^{(b)}$.

Step 2: Calculate cluster centres $v(b)$ with Eq. (5.7) based on $U^{(b)}$.

Step 3: Update $U^{(b+1)}$ with cluster centres $v(b)$ by Eq. (5.8).

Step 4: If $\left\| U^{(b+1)} - U^{(b)} \right\| \leq \varepsilon$, terminate the iterations and obtain cluster centres v and the FCM partition matrix U. The clustering process stops when the maximum number of iterations is reached, or when the objective function improvement between two consecutive iterations is less than the minimum amount of improvement specified. Otherwise, set $b = b + 1$ and return to step 2:

$$v_j^{(b)} = \left[\sum_{i=1}^{N} (u_{ij}^{(b)})^m \cdot x_i \right] \Big/ \left[\sum_{i=1}^{N} (u_{ij}^{(b)})^m \right], \quad j = 1, 2, \ldots, c \tag{5.7}$$

$$u_{ij}^{(b+1)} = \left[\sum_{k=1}^{c} (d(x_i, v_j^{(b)}) \Big/ d(x_i, v_k^{(b)}))^{2/(m-1)} \right]^{-1}, \quad k = 1, 2, \ldots, c \tag{5.8}$$

5.4. Experiments and Results

5.4.1. *Experimental setup*

In consultation with experts from Joy Global, we understand that anomalies would occur repeatedly for a period of time (e.g. several days in a row) before a mining machine breaks down. Ideally, we should alert technicians to check the machine prior to its breakdown. This means an alarm should be triggered if persistent anomalies are observed. Given that the original grease cycle data was recorded with frequencies ranging from 3 to 30 minutes, we have identified two ways to detect anomalies:

Case (1): We use the raw data as input to train clustering models. Anomalies at the minute level are first detected based on the models. Then, we label a day as alarming if the number of anomalies detected in that day is more than a certain percentage, denoted as p, over the sample total in the same day.

Table 5.2. Statistics of the first 10 samples from the grease cycle dataset.

Minimum	Maximum	Mean	Standard deviation
12	25.4	16.43498	3.30282
13.9	144.2	20.52158	10.14567
13.9	28.4	20.07933	3.466548
12.9	27.4	18.22584	3.538307
14.8	25.9	19.58291	3.241977
12	26.9	18.51689	3.672978
14.9	29.8	20.82667	3.534569
10.1	21.2	16.1457	2.783742
9	16.9	13.21429	2.031048
11.1	20.6	15.54803	2.528468

Case (2): Based on the original dataset, we calculate the minimum, maximum, mean and standard deviation of each day's data. Records with only one point in a day are deleted to avoid the standard deviation being zero. Examples of extracted daily data are shown in Table 5.2. The daily statistics are used as input to train clustering models. Daily anomalies are directly labelled based on the clustering results.

In all our experiments, data from 2015 was used as the training set, and the rest for testing.

5.4.2. *Parameter settings*

For each clustering algorithm, the number of clusters was set to 2, corresponding to normal and abnormal groups. The distance measure used was the squared Euclidean distance, and the number of iterations was set at 100. For k-means, initial cluster centres were randomly selected from the samples. For FCM, the minimum amount of improvement was e^{-5}, and the exponent for matrix U was set to 2 (which is the value of m in Eq. (5.6)). For the SOM, the Kohonen learning rate was set to 0.1, and the conscience learning rate was set to 0.001. Performance in training was measured using the mean-squared error.

The percentage of daily anomalies, p, was set to 0.6 in this study. This means that if the number of anomalies detected in a day is more than 60% of the sample total in that day, the day is flagged as alarming.

5.4.3. *Clustering results*

Figures 5.4–5.6 show the clustering results of k-means, FCM and the SOM, respectively, based on Case (1) as discussed in Section 5.4.1. From these figures, we see that the clustering models divide the data into two groups. We take those with large fluctuations as abnormal data, and the others as normal data. As shown in these figures, the number of anomalies varies from 20 to 180.

Table 5.3 shows the specific number of normal and abnormal samples detected by k-means, FCM and the SOM based on the raw data.

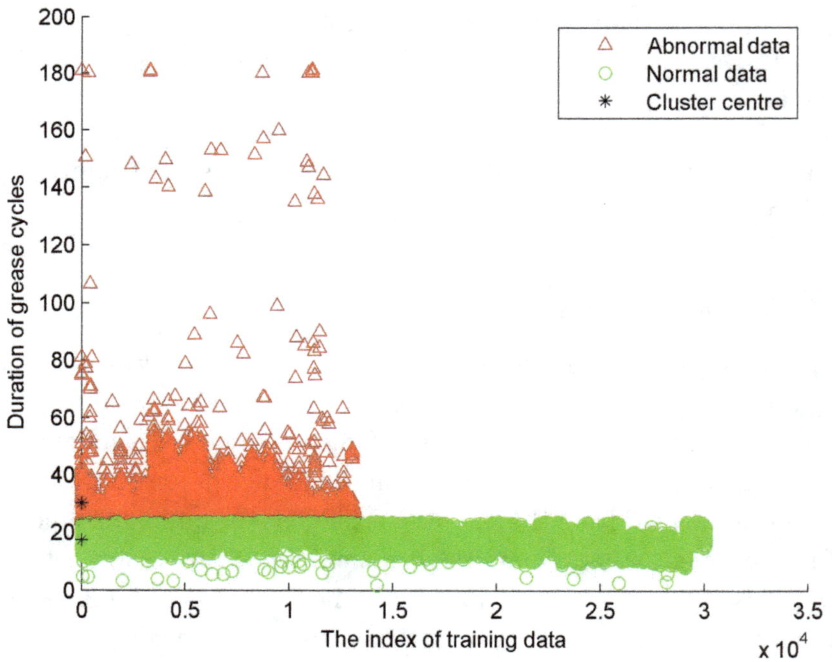

Fig. 5.4. Clustering results using k-means.

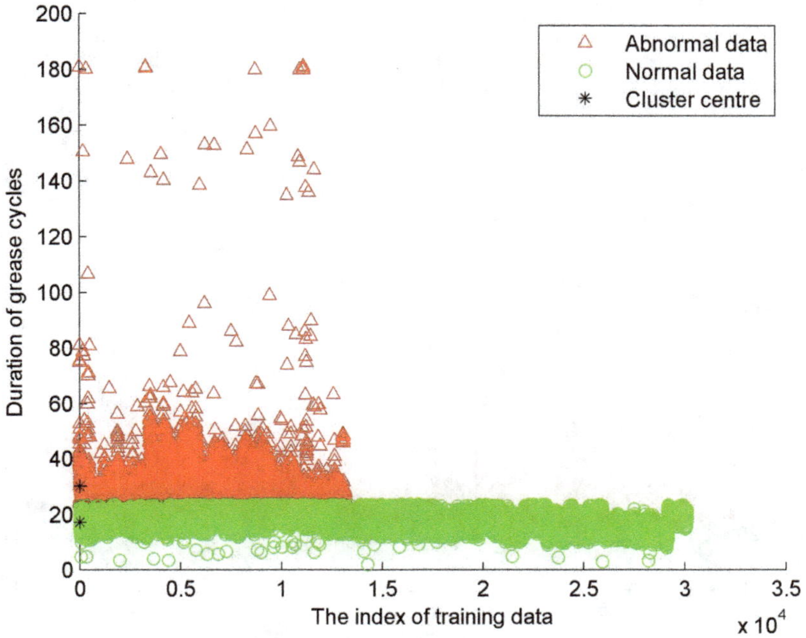

Fig. 5.5. Clustering results using FCM.

From this table, we observe that more than one-third of the samples have been labelled as abnormal by the three algorithms in both training and testing data. Many of these anomalies could have been caused by external factors such as temperature and moisture and cannot be taken seriously. It is also worth noting that these are anomalies at the minute level.

As per Case (1), we treat a day as alarming if the number of anomalies exceeded the 60% threshold in that day. Table 5.4 shows the number of days detected as alarming based on results from Table 5.3. As can be seen in Table 5.4, more than half of the days have been flagged as alarming by k-means in both the training and testing data. These results are not useful at all. FCM has presented a more reasonable solution (96 days) with the testing data but not the training data (175 days). Overall, the SOM produces the most reasonable solutions (73 days with training data and 111 days with testing data)

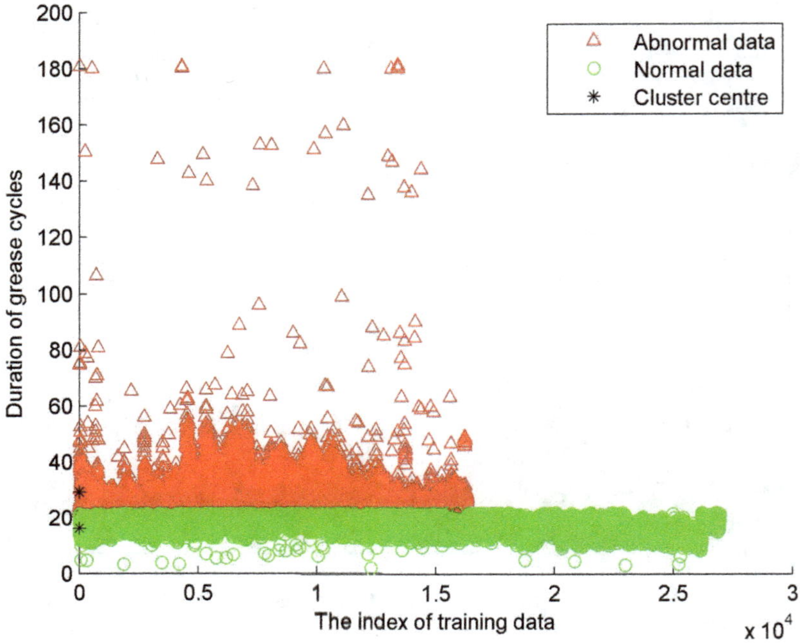

Fig. 5.6. Clustering results using the SOM algorithm.

Table 5.3. Anomalies detected by k-means, FCM and the SOM based on the raw data.

	Algorithm	k-means	FCM	SOM
Training	Normal data	30096	30090	28607
	Abnormal data	13155	13161	14644
Testing	Normal data	25707	25425	23882
	Abnormal data	15702	15984	17527

Table 5.4. Daily anomalies detected based on results from Table 5.3.

	Algorithm	k-means	FCM	SOM
Training	Normal data	121	121	223
	Abnormal data	175	175	73
Testing	Normal data	191	257	242
	Abnormal data	162	96	111

Table 5.5. Daily anomalies detected based on statistical measures.

	Algorithm	k-means	FCM	SOM
Training	Normal data	268	267	249
	Abnormal data	26	27	45
Testing	Normal data	331	298	305
	Abnormal data	20	53	46

among the three clustering models, but the results imply that technicians would need to check the machines regularly for close to one-third of a year.

Using extracted daily data as per Case (2), results obtained by the three clustering models are shown in Table 5.5. Compared to the results of Case (1) in Table 5.4, there are much fewer alarming days in both training and testing data, indicating that the number of false alarms is largely reduced. While there are only 20–27 days flagged by k-means and FCM, with the exception of 53 days for FCM in the testing data, the SOM has identified a more conservative outcome of around 45–46 days. Checking the machines around 45–46 days appears to be acceptable and a safer approach.

5.4.4. *Discussion*

Figure 5.7 shows an example of real-world time series data provided by Joy Global over a period of three months. We can see from the figure that there are two spikes during the period. The first spike, however, was not of concern as it could be something caused by random factors such as a sudden change of climate or temperature. The second spike led to a machine breakdown and required technicians to repair the machine. Before the second spike, we observe that there is a step-up, which persisted until the ramp-up occurs. After the repair, we see that the time series goes back to normal.

From Fig. 5.7, we learn that an outlier or anomaly such as the first spike, even though it looks exactly like the second spike, can be a false alarm. Delving further into other similar time series data from Joy Global (not shown here), we confirm that a machine breakdown

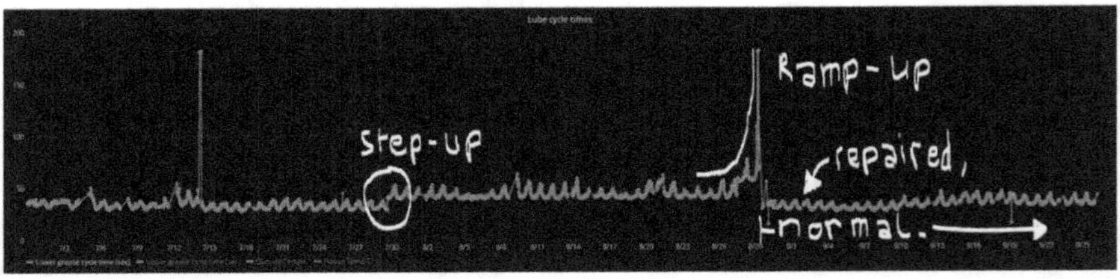

Fig. 5.7. A screenshot of real-world time series data leading to machine breakdown.

typically occurs after a consistent period of step-up (e.g. 12–15 days, or even close to a month like the scenario shown in Fig. 5.7). In other words, we should not be alarmed by anomalies that do not persist for days. Rather than alerting the technicians of anomalies based on the p threshold as per Section 5.4.3, we would like to recommend a very simple but more time-saving way for Joy Global to prevent machine failures: the alarm will be triggered if anomalies detected persist for n days in a row.

To demonstrate the idea, we used the same dataset and same clustering models but this time anomalies will be flagged only if they happened seven days in a row (i.e. $n = 7$). Results obtained by k-means, FCM and the SOM are shown in Figs. 5.8–5.10, respectively, where alarms triggered are represented by red asterisks. The 0 value of y-axis corresponds to normal data, while 1 denotes anomalies. From the figures, we see that the alarm is triggered only three times by k-means and FCM, while other anomalies are ignored. In fact, these three triggered cases refer to the same incident (the alarm will be triggered repeatedly every 7 days, if no one attends to the machine). As for the SOM, the alarm is triggered six times, with three separate incidents identified. This shows that the SOM is more reliable than k-means and FCM, as we do not want to miss the chance of fixing potential issues.

It is worth pointing out that the value of n is adjustable depending on the company's situation as well as conditions of the machines. If major failures occur within the seven-day range, the value of n can be reduced to ensure more regular checks are carried out.

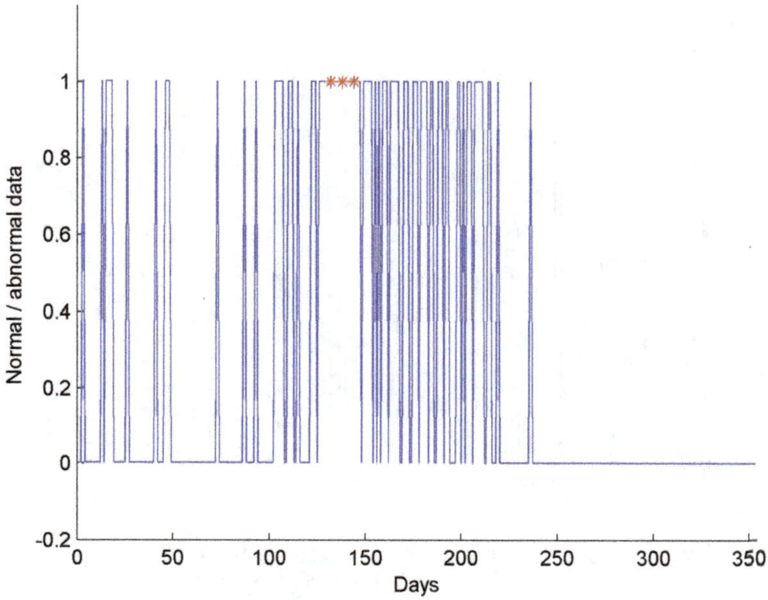

Fig. 5.8. Alarms triggered by k-means based on the testing data.

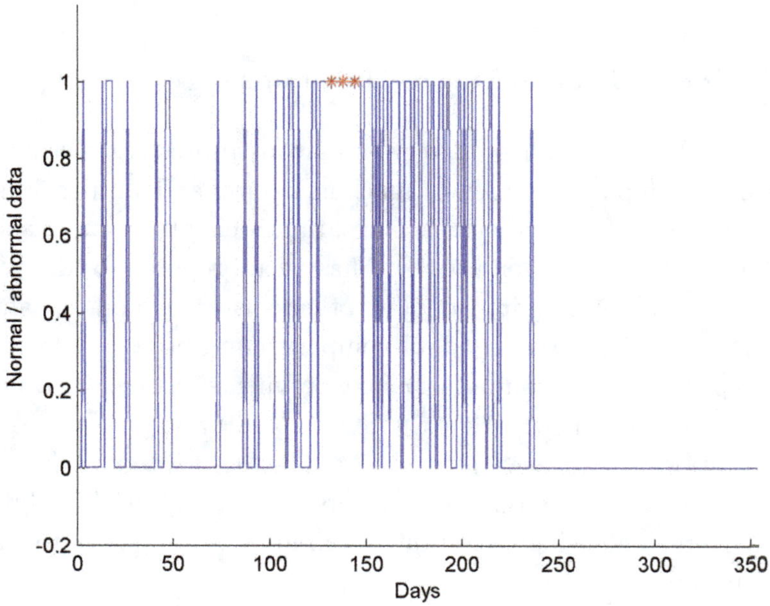

Fig. 5.9. Alarms triggered by FCM based on the testing data.

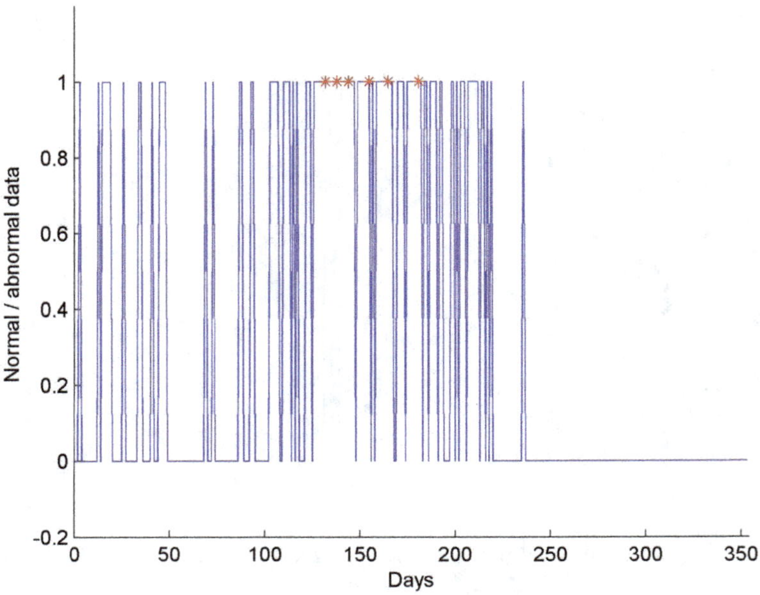

Fig. 5.10. Alarms triggered by the SOM based on the testing data.

5.5. Conclusion and Future Directions

Anomaly detection can be a very important protecting mechanism for mining companies to prevent machine failures and, in turn, reduce the cost of maintaining their machines. In this chapter, a bio-inspired clustering model based on the SOM has been proposed for detecting anomalies in mining equipment. Two other clustering models, namely k-means and FCM, were used for comparison purposes to assess the SOM's performance. Computational experiments based on real-world grease cycle data showed that the SOM could provide more reasonable solutions than k-means and FCM. The SOM was also shown to be more capable of identifying incidents of potential machine breakdown, thereby allowing technicians from mining companies to check their machines.

The work presented in this chapter was our preliminary collaboration with Joy Global. In the next stage, we hope to further improve the proposed model by optimizing its parameters and structures using some bio-inspired optimization algorithms [6, 7]. We will also consider more advanced methods, including probability distributions, fuzzy-based models and deep-learning models, to detect anomalies in mining machines, should more real-world data become available. While it is very time-consuming, we plan to label anomalies in sensor data provided with the help of experts from Joy Global to better evaluate our anomaly detection models. The ultimate goal is to incorporate these models in practical applications within the mining industry for preventing machine failures.

Bibliography

[1] Aggarwal, C. C. Data mining. In *Outlier Analysis*, Springer International Publishing, pp. 237–263, 2015.
[2] Ben-Hur, A. *et al.* Support vector clustering. *J. Mach. Learn. Res.* **2**, 125–137, 2001.
[3] Boecking, B. *et al.* Support vector clustering of time series data with alignment kernels. *Pattern Recognit. Lett.* **45**, 129–135, 2014.
[4] Brandt, T., Grawunder, M. and Appelrath, H. J. Anomaly detection on data streams for machine condition monitoring. In *2016 IEEE 14th International Conference on Industrial Informatics (INDIN)*, pp. 1282–1287, 2016.
[5] Chandola, V., Banerjee, A. and Kumar, V. Anomaly detection: A survey. *ACM Comput. Surv.* **41**(3), 15, 2009.
[6] Chiong, R. and Weise, T. *Variants of Evolutionary Algorithms for Real-World Applications.* Springer, 2012.
[7] Chiong, R. *Nature-Inspired Algorithms for Optimisation.* Springer, 2009.
[8] Dunn, J. C. A fuzzy relative of the ISODATA process and its use in detecting compact well-separated clusters. *J. Cybern.* **3**(3), 32–57, 1973.
[9] Erfani, S. M. *et al.* High-dimensional and large-scale anomaly detection using a linear one-class SVM with deep learning. *Pattern Recognit.* **58**, 121–134, 2016.
[10] Frey, C. W. Diagnosis and monitoring of complex industrial processes based on self-organizing maps and watershed transformations. In *2008 IEEE International Conference on Computational Intelligence for Measurement Systems and Applications (CIMSA)*, pp. 87–92, 2008.
[11] Girdhar, Y. *et al.* Anomaly detection in unstructured environments using bayesian nonparametric scene modelling. In *2016 IEEE International Conference on Robotics and Automation (ICRA)*, pp. 2651–2656, 2016.

[12] Hall, R. A. and Daneshmend, L. K. Reliability modelling of surface mining equipment: data gathering and analysis methodologies. *Int. J. Surf. Mining Reclamation Environ.* **17**(3), 139–155, 2003.
[13] Han, J., Pei, J. and Kamber, M. *Data Mining: Concepts and Techniques.* Elsevier, 2011.
[14] Hastie, T., Tibshirani, R. and Friedman, J. *The Elements of Statistical Learning: Data Mining, Inference, and Prediction.* 2nd edn. Springer, New York, 2009.
[15] Jain, A. K. Data clustering: 50 years beyond k-means. *Pattern Recognit. Lett.* **31**(8), 651–666, 2010.
[16] Jalalian, A. and Chalup, S. K. GDTW-P-SVMs: Variable-length time series analysis using support vector machines. *Neurocomputing* **99**, 270–282, 2013.
[17] Kohonen, T. Essentials of the self-organizing map. *Neural Netw.* **37**, 52–65, 2013.
[18] Lee, J. and Lee, D. An improved cluster labeling method for support vector clustering. *IEEE Trans. Pattern Anal. Mach. Intell.* **27**(3), 461–464, 2005.
[19] Lei, J. Z. and Ghorbani, A. Network intrusion detection using an improved competitive learning neural network. In *Proceedings. Second Annual Conference on Communication Networks and Services Research*, pp. 190–197, 2004.
[20] Lewis, M. W. and Steinberg, L. Maintenance of mobile mine equipment in the information age. *J. Qual. Maint. Eng.* **7**(4), 264–274, 2001.
[21] Möller-Levet, C. S. *et al.* Fuzzy clustering of short time-series and unevenly distributed sampling points. In *International Symposium on Intelligent Data Analysis.* Springer, Berlin, pp. 330–340, 2003.
[22] Murtagh, F. A survey of recent advances in hierarchical clustering algorithms. *Comput. J.* **26**(4), 354–359, 1983.
[23] Ogawa, H. *et al.* Experimental fault analysis process implemented using model extraction and model checking. In *2015 IEEE 39th Annual Computer Software and Applications Conference (COMPSAC)*, Vol. 2, pp. 95–104, 2015.
[24] Oliveira, J. V. D. and Pedrycz, W. *Advances in Fuzzy Clustering and its Applications.* John Wiley & Sons, Chichester, 2007.
[25] Pal, N. R. and Bezdek, J. C. On cluster validity for the fuzzy c-means model. *IEEE Trans. Fuzzy Syst.* **3**(3), 370–379, 1995.
[26] Pham, T., Dang, H., Le, T. and Le, T. H. Fast support vector clustering. *Vietnam J. Comput. Sci.* **4**(1), 13–21, 2017.
[27] Saltos, R. and Weber, R. A rough–fuzzy approach for support vector clustering. *Inf. Sci.* **339**, 353–368, 2016.
[28] Samantaray, S. R. Genetic-fuzzy rule mining approach and evaluation of feature selection techniques for anomaly intrusion detection. *Appl. Soft Comput.* **13**, 928–938, 2013.
[29] Tsai, W. P. *et al.* A data-mining framework for exploring the multi-relation between fish species and water quality through self-organizing map. *Sci. Total Environ.* **579**, 474–483, 2017.
[30] Vesanto, J. and Alhoniemi, E. Clustering of the self-organizing map. *IEEE Trans. Neural Netw.* **11**(3), 586–600, 2000.

[31] Yang, J., Estivill-Castro, V. and Chalup, S. K. Support vector clustering through proximity graph modelling. In *Proceedings of the 9th International Conference on Neural Information Processing (ICONIP'02)*, Vol. 2, pp. 898–903, 2002.

[32] Yassin, W. *et al.* Anomaly-based intrusion detection through k-means clustering and naives Bayes classification. In *Proceedings of the 4th International Conference on Computing and Informatics (ICOCI)*, Paper no. 49, pp. 298–303, 2013.

[33] Yu, J., Yang, M. and Lee, E. S. Sample-weighted clustering methods. *Comput. Math. Appl.* **62**(5), 2200–2208, 2011.

[34] Yu, J. and Yang, M. Optimality test for generalized FCM and its application to parameter selection. *IEEE Trans. Fuzzy Syst.* **13**(1), 164–176, 2005.

[35] Zimek, A., Schubert, E. and Kriegel, H. P. A survey on unsupervised outlier detection in high-dimensional numerical data. *Stat. Anal. Data Min.* **5**(5), 363–387, 2012.

Chapter 6

Picture Generating Models Based on the Splicing Operation

K. G. Subramanian*, G. Samdanielthompson and N. Gnanamalar David

Department of Mathematics, Madras Christian College
Tambaram, Chennai 600059, India
**kgsmani1948@gmail.com*

The operation of splicing on words, introduced by Head [Formal language theory and DNA: An analysis of the generative capacity of specific recombinant behaviors. *Bull. Math. Biol.* **49**, 735, 1987] in his pioneering work on an analysis of the recombinant behaviour of DNA molecules, has been instrumental for a series of language-theoretic studies related to both string languages and picture array languages. Recently, a specific kind of splicing, called flat splicing on words was introduced. Subsequently, flat splicing on picture arrays was considered and two-dimensional (2D) picture array generating grammars and grammar systems have been proposed. On the other hand, pure 2D context-free grammar ($P2DCFG$), recently introduced, is a simple yet expressive 2D array generating grammar. This chapter first briefly reviews the picture array generating models based on the picture array splicing operation and in particular array flat splicing operation. Then a new 2D picture array generating grammar system with its components involving $P2DCFG$ kind of rules and the components "communicating" through array flat splicing operation is introduced and its generative power is examined. The applicability of this new system in the generation of picture patterns, commonly referred to as "floor designs", is also exhibited.

6.1. Introduction

The birth of the area of DNA computing [9, 12, 16, 17] could be attributed in some sense to Adleman's exciting announcement [1] of solving a small instance of the Hamiltonian path problem based on the use of DNA molecules and operations on DNA sequences. A new direction of research and study resulted in DNA computing beginning with the seminal work of Head [11, 12] on modelling the recombinant behaviour of DNA molecules under the influence of restriction enzymes and ligase. The study of Head [11] was in the framework of formal language theory [8, 14, 20], which is considered to be an important branch of theoretical computer science. Head [11] introduced a formal model of DNA recombinant behaviour, called a splicing system, based on a new language-theoretic operation of splicing words, which stimulated great interest in researchers and opened up investigations resulting in deep and interesting results [16] in the area of formal language theory.

On the other hand, motivated by different problems, such as the formation of floor patterns (also called "kolam" designs) [2, 23], tiling a plane area, generating geometric shapes like squares and rectangles and many others, a variety of two-dimensional (2D) picture array generating grammar models [10, 18, 19, 25] have been introduced and investigated by researchers, basically based on the concepts and rich results in the area of formal string languages [20]. We are mainly concerned here with non-isometric classes of 2D grammars [19, 23, 25], which, unlike the isometric array grammars [18], can alter the geometric shape of a picture array while rewriting is done by the rules of grammar. Pure 2D context-free grammar ($P2DCFG$) [25] is a recently introduced non-isometric variety of 2D grammar, which is simple to handle yet expressive [3] in power for picture array generation.

With formal language theory as a framework for modelling distributed complex systems, the notion of a grammar system [6] in which a collection of grammars work together as specified by a well-defined protocol was introduced. Among a variety of grammar systems, splicing grammar system was introduced in [7] with the component grammars working in parallel and communication between components being done by the bio-inspired operation of splicing on strings of symbols. The concept of splicing on words was extended [5] to rectangular picture arrays. Based on this array splicing operation, splicing array grammar system [26] with the components involving regular or context-free array rewriting rules and the array splicing operation was introduced. Recently, a specific kind of splicing on words, called flat splicing, was introduced [4]. This notion was also extended to picture arrays and a picture array generating system based on array flat splicing operation was introduced [27]. Also, a grammar system based on regular or context-free array rewriting rules and the array flat splicing operation was also considered [21].

We first briefly and informally review, in the subsequent sections, the notions of the picture array generating systems based on the array splicing operation. Then a new picture array generating grammar system, called flat splicing pure 2D context-free array grammar system ($FSP2DCFGS$), which has $P2DCFG$ type of rules in its components and which "communicates" among its components by array flat splicing operation is introduced. The resulting class of picture array languages is compared with certain other picture array generating models. As an application, an $FSP2DCFGS$ is constructed to generate a picture language whose elements correspond to certain "floor designs", also known as "kolam patterns" [2].

6.2. Preliminaries

We recall a few needed notions of formal language theory, array grammars and 2D languages [10, 18, 20].

A word w over an alphabet Σ is a string of symbols of the form $w = a_1a_2 \cdots a_n$, $a_i \in \Sigma$, $1 \leq i \leq n$. The length of a word w is denoted by $|w|$. The set of all words over an alphabet Σ is denoted by Σ^*, which includes the empty word λ with no symbols. For any

$w = a_1 a_2 \cdots a_n$, we denote by ${}^t w$ the vertical word

$$\begin{matrix} a_1 \\ a_2 \\ \vdots \\ a_n \end{matrix}$$

We also define ${}^t({}^t w) = w$. We set ${}^t\lambda$ as λ itself.

A rectangular $m \times n$ array M over Σ (also called a picture array) is of the form

$$M = \begin{matrix} a_{11} & \cdots & a_{1n} \\ \vdots & \ddots & \vdots \\ a_{m1} & \cdots & a_{mn} \end{matrix}$$

where each $a_{ij} \in \Sigma, 1 \leq i \leq m, 1 \leq j \leq n$. The set of all rectangular arrays over Σ is denoted by Σ^{**}, which includes the empty array λ. $\Sigma^{++} = \Sigma^{**} - \lambda$. We denote, respectively, by $\circ$ and $\diamond$ the *column concatenation* and *row concatenation* of arrays in Σ^{**}. These array operations are partially defined, namely for any A, $B \in \Sigma^{**}$, $A \circ B$ is defined if and only if A and B have the same number of rows. Similarly $A \diamond B$ is defined if and only if A and B have the same number of columns. For example, if

$$A = \begin{matrix} a\,b\,a\,b \\ b\,b\,a\,a \\ a\,b\,b\,a \end{matrix}, \quad B = \begin{matrix} a\,a\,a\,b \\ b\,b\,a\,b \\ a\,b\,b\,b \end{matrix}$$

are two 3×4 arrays over an alphabet $\{a, b\}$, then

$$A \circ B = \begin{matrix} a\,b\,a\,b\,a\,a\,a\,b \\ b\,b\,a\,a\,b\,b\,a\,b \\ a\,b\,b\,a\,a\,b\,b\,b \end{matrix}$$

Likewise, $A \diamond B$ is formed placing B below A.

We now recall pure context-free string grammars [15] which, unlike a context-free grammar, has only one kind of a symbol, namely the terminal symbol.

A pure context-free grammar [15] is $G = (\Sigma, P, \mathbb{A})$, where Σ is a finite alphabet, $\mathbb{A}$ is a set of axiom words and P is a finite set of context-free rules of the form $a \to \alpha, a \in \Sigma, \alpha \in \Sigma^*$. Derivations are done as in context-free grammar. The language generated consists of all words derived from each axiom word. For example, the pure context-free grammar $G = (\{a, b, e\}, P = \{e \to aeb, \}, \{aeb\})$ generates the language $\{a^n eb^n \mid n \geq 1\}$. We next recall the operations of splicing [11, 16] and flat splicing on words [4].

The operation of splicing applied on two words $u = u_1\alpha\beta u_2$ and $v = v_1\gamma\delta v_2$ over an alphabet Σ involves two steps: (i) the word u is "cut" between α and β, and v between γ and δ according to a splicing rule of the form $(\alpha, \beta; \gamma, \delta)$ and (ii) the prefix $u_1\alpha$ is joined with the suffix δv_2 yielding a word $u_1\alpha\delta v_2$.

A flat splicing rule r is of the form $(\alpha|\gamma - \delta|\beta)$, where $\alpha, \beta, \gamma, \delta$ are words over a given alphabet Σ. For two words $u = x\alpha\beta y$, $v = \gamma z\delta$, where x, y, z are words over Σ, an application of the flat splicing rule r to the pair (u, v) yields the word $w = x\alpha\gamma z\delta\beta y$. In other words, the second word v is inserted between α and β in the first word u as a result of applying the rule r.

6.3. Splicing Array Grammar Systems and Picture Languages

We first discuss 2D grammar generating rectangular picture arrays.

A 2D context-free grammar ($2CFG$) (in the terminology used in [10] and initially called as context-free matrix grammar [23]) is a rectangular picture array generating a grammar model. This involves two phases of derivation. In the first phase, a string of intermediate symbols is generated from a start symbol by context-free string grammar rules. In the second phase, all the intermediate symbols are rewritten in the vertical direction in parallel by right linear grammar rules, such that either all the rules used at a time are non-terminal rules of the form $X \to AY$ or all the rules used are terminal rules of the form $X \to A$ where X, Y are non-terminals and A is an intermediate symbol, with the derivation terminating on applying terminal rules and yielding a rectangular picture array. If the rules in the first phase are regular grammar rules, then the grammar is called a 2D regular grammar ($2RLG$) (initially called as regular matrix grammar [23]).

An extension of $2CFG$ was introduced [26] by considering the second phase groups of regular non-terminal rules or groups of regular terminal rules, referred to as "tables" of rules and at a step in the derivation in the vertical direction, rules of a table are applied. The resulting grammar is referred to as tabled $2CFG$ or tabled $2RLG$ ($T2CFG$ or $T2RLG$) depending on the first phase involving context-free or regular rules.

We illustrate with an example. We state only the rules instead of formally describing the components. Consider the $T2RLG$ with the first phase having the rules $S \to AX, X \to BX, X \to BY, Y \to A$, where S, X, Y are non-terminals with S as the start symbol and A, B are intermediates. The second phase has the following tables of rules $t_1 = \{A \to aA, B \to bB\}$, $t_2 = \{A \to aC, B \to aD\}$, $t_3 = \{C \to aC, D \to bD\}$, $t_4 = \{C \to a, D \to b\}$, where a, b are terminals. The first phase generates words of the form AB^nA, $n \geq 1$. In the second phase, for example, a word of the form $ABBA$ generates an array representing the letter H by interpreting b as blank. The sequence of steps, that uses the tables, t_1, t_1, t_2, t_3, t_4, is as follows:

$$A\,B\,B\,A \Rightarrow \begin{matrix} a & b & b & a \\ A & B & B & A \end{matrix} \Rightarrow \begin{matrix} a & b & b & a \\ a & b & b & a \\ A & B & B & A \end{matrix} \Rightarrow \begin{matrix} a & b & b & a \\ a & b & b & a \\ a & a & a & a \\ C & D & D & C \end{matrix}$$

$$\Rightarrow \begin{matrix} a & b & b & a \\ a & b & b & a \\ a & a & a & a \\ a & b & b & a \\ C & D & D & C \end{matrix} \Rightarrow \begin{matrix} a & b & b & a \\ a & b & b & a \\ a & a & a & a \\ a & b & b & a \\ a & b & b & a \end{matrix}$$

We now deal with splicing on arrays. The splicing operation has been extended to rectangular arrays in [5]. In informal terms, we recall this notion. This is different from the way array splicing is done in [13].

Given two rectangular arrays A, B with the same number of rows, in column array splicing, each of the arrays A, B is "cut" between two specified columns so that A is "split" into a left part A_1 and a right part A_2 giving $A = A_1A_2$ while B is "split" into a left part B_1 and a right part B_2 giving $B = B_1B_2$. The array A_1 is then "pasted" (or column catenated) with B_2 yielding the array A_1B_2. The "cut between columns is decided by a finite sequence of rules of the form $r : \left(\begin{smallmatrix} a_1 \\ a_2 \end{smallmatrix}, \begin{smallmatrix} b_1 \\ b_2 \end{smallmatrix}; \begin{smallmatrix} c_1 \\ c_2 \end{smallmatrix}, \begin{smallmatrix} d_1 \\ d_2 \end{smallmatrix}\right)$. Informally expressed, a 2×2 "window" is moved vertically down on two adjacent columns (one of which could be empty) from the first to the last row in each of the two arrays A, B at the same time and at any step, if the first and the second columns of the window on every two consecutive rows of A are the dominoes $\begin{smallmatrix} a_1 \\ a_2 \end{smallmatrix}, \begin{smallmatrix} b_1 \\ b_2 \end{smallmatrix}$ and in the corresponding window on B are the dominoes $\begin{smallmatrix} c_1 \\ c_2 \end{smallmatrix}, \begin{smallmatrix} d_1 \\ d_2 \end{smallmatrix}$ such that the dominoes constitute a column array splicing rule of the form r, then the "cut" can be affected. Analogously, we can consider splicing on arrays with the "cut" and "paste" operations on rows.

Motivated by the study of Dassow and Mitrana [7], a grammar system, called splicing array grammar system ($SAGS$), that describes pictures of rectangular arrays was introduced [26]. The components of this grammar system consist of $T2CFG$ or $T2RLG$ types of rules and the system also has domino column or row splicing rules [26] with the grammars working in parallel and splicing rules acting on arrays of two components yielding rectangular arrays of symbols.

We illustrate with an example.

Example 6.1. Let $V = \{a, a_1, a_2, b, b_1, b_2\}$ be the set of symbols and

$$I = \left\{\begin{matrix} a_2\, b\, a\, b_1 \\ b_2\, a\, b\, a_1 \\ a_2\, b\, a\, b_1 \\ b_2\, a\, b\, a_1 \end{matrix}\right\}$$

be the starting array. $R_c = \left\{p_1 : \left(\begin{smallmatrix} b_1 \\ a_1 \end{smallmatrix}, \lambda;\ \begin{smallmatrix} a_2 \\ b_2 \end{smallmatrix}, \lambda\right), p_2 : \left(\begin{smallmatrix} a_1 \\ b_1 \end{smallmatrix}, \lambda;\ \begin{smallmatrix} b_2 \\ a_2 \end{smallmatrix}, \lambda\right)\right\}$ is the set of rules with p_1 and p_2 two splicing rules.

Fig. 6.1. A chessboard pattern.

One column array splicing rule is used in a parallel manner.

$$\begin{matrix} a_2 & b & a & b_1 \\ b_2 & a & b & a_1 \\ a_2 & b & a & b_1 \\ b_2 & a & b & a_1 \end{matrix}$$

By using the rules of p_1 and p_2 in parallel, we obtain the array

$$\begin{matrix} a_2 & b & a & b_1 & a_2 & b & a & b_1 \\ b_2 & a & b & a_1 & b_2 & a & b & a_1 \\ a_2 & b & a & b_1 & a_2 & b & a & b_1 \\ b_2 & a & b & a_1 & b_2 & a & b & a_1 \end{matrix}$$

On mapping a, a_1 and b, b_1, respectively, to black and white coloured unit square cells, we get a chess board pattern as shown in Fig. 6.1.

6.4. Array Flat Splicing Systems

The notion of flat splicing on words [4] has been extended to arrays by [27] introducing column and row flat splicing rules and thus a new model of picture generation called array flat splicing system is introduced [27]. We now recall the operation of flat splicing on arrays.

Definition 6.1 ([27]). Let V be an alphabet.

(1) A column flat splicing rule is of the form $({}^t(a_1a_2)|{}^t(x_1x_2) - {}^t(y_1y_2)|{}^t(b_1b_2))$, where $a_1, a_2, b_1, b_2 \in \Sigma \cup \{\lambda\}$ with $|a_1| = |a_2|$

and $|b_1| = |b_2|$, $x_1, x_2, y_1, y_2 \in \Sigma \cup \{\lambda\}$ with $|x_1| = |x_2|$ and $|y_1| = |y_2|$.

(2) A row flat splicing rule is of the form $(c_1c_2|u_1u_2 - v_1v_2|d_1d_2)$, where $c_1, c_2, d_1, d_2 \in \Sigma \cup \{\lambda\}$ with $|c_1| = |c_2|$ and $|d_1| = |d_2|$, $u_1, u_2, v_1, v_2 \in \Sigma \cup \{\lambda\}$ with $|u_1| = |u_2|$ and $|v_1| = |v_2|$.

(3) Let $r_1, r_2, \ldots, r_{m-1}$ be a sequence of $(m-1)$ column flat splicing rules given by

$$r_i = ({}^t(\alpha_i\alpha_{i+1})|^t(\gamma_i\gamma_{i+1}) - {}^t(\delta_i\delta_{i+1})|^t(\beta_i\beta_{i+1}))$$

for $1 \le i \le (m-1)$. Let X, Y be two picture arrays, each with m rows, for some $m \ge 1$, and given by

$$X = X_1 \circ {}^t(\alpha_1\alpha_2\cdots\alpha_m) \circ {}^t(\beta_1\beta_2\cdots\beta_m) \circ X_2$$
$$Y = {}^t(\gamma_1\gamma_2\cdots\gamma_m) \circ Y' \circ {}^t(\delta_1\delta_2\cdots\delta_m)$$

where X_1, X_2, Y' are arrays over Σ with m rows, $\alpha_i, \beta_i \in \Sigma\cup\{\lambda\}$ $(1 \le i \le m)$, with $|\alpha_1| = |\alpha_2| = \cdots = |\alpha_m|$, $|\beta_1| = |\beta_2| = \cdots = |\beta_m|$, γ_i, δ_i, $(1 \le i \le m) \in \Sigma \cup \{\lambda\}$ with $|\gamma_1| = |\gamma_2| = \cdots = |\gamma_m|$, $|\delta_1| = |\delta_2| = \cdots = |\delta_m|$. An application of the column flat splicing rules $r_1, r_2, \ldots, r_{m-1}$ to the pair of arrays (X, Y) yields the array

$$\begin{aligned} Z &= X_1 \circ {}^t(\alpha_1\alpha_2\cdots\alpha_m) \circ {}^t(\gamma_1\gamma_2\cdots\gamma_m) \\ &\quad \circ Y' \circ {}^t(\delta_1\delta_2\cdots\delta_m) \circ {}^t(\beta_1\beta_2\cdots\beta_m) \circ X_2 \end{aligned}$$

The pair (X, Y) yielding Z is then denoted by $(X, Y) \vdash_c Z$.

(4) Let $s_1, s_2, \ldots, s_{n-1}$ be a sequence of $(n-1)$ row flat splicing rules given by

$$s_j = (\eta_j\eta_{j+1}|(\mu_j\mu_{j+1}) - (\nu_j\nu_{j+1})|\theta_j\theta_{j+1})$$

for $1 \le j \le (n-1)$. Let U, V be two picture arrays, each with n columns, for some $n \ge 1$, and given by

$$U = U_1 \diamond (\eta_1\eta_2\cdots\eta_n) \diamond (\theta_1\theta_2\cdots\theta_n) \diamond U_2$$
$$V = (\mu_1\mu_2\cdots\mu_n) \diamond V' \diamond (\nu_1\nu_2\cdots\nu_n)$$

where U_1, U_2, V' are arrays over Σ with n columns, η_j, θ_j, $(1 \le j \le n) \in \Sigma \cup \{\lambda\}$ with $|\eta_1| = |\eta_2| = \cdots = |\eta_n|$, $|\theta_1| = |\theta_2| =$

$\cdots = |\theta_n|$, μ_j, ν_j, $(1 \leq j \leq n) \in \Sigma \cup \{\lambda\}$ with $|\mu_1| = |\mu_2| = \cdots = |\mu_n|$, $|\nu_1| = |\nu_2| = \cdots = |\nu_n|$. An application of the row flat splicing rules $s_1, s_2, \ldots, s_{n-1}$ to the pair of arrays (U, V) yields the array W

$$= U_1 \diamond (\eta_1\eta_2 \cdots \eta_n) \diamond (\mu_1\mu_2 \cdots \mu_n) \diamond V' \diamond (\nu_1\nu_2 \cdots \nu_n) \\ \diamond (\theta_1\theta_2 \cdots \theta_n) \diamond U_2$$

The pair (U, V) yielding W is then denoted by $(U, V) \vdash_r W$.

(5) An array flat splicing rule is either a column flat splicing rule or a row flat splicing rule. The notation $\vdash$ denotes either $\vdash_c$ or $\vdash_r$.

(6) For a picture language $L \subseteq \Sigma^{**}$ and a set R of array flat splicing rules, we define

$$f(L) = \{M \in \Sigma^{**} \mid (X, Y) \vdash M, \text{for } X, Y \in L, \\ \text{and some rule in } R\}$$

(7) An array flat splicing system (AFS) is $\mathcal{A} = (\Sigma, M, R_c, R_r)$, where Σ is an alphabet, M is a finite set of arrays over Σ, called initial set, R_c is a finite set of column flat splicing rules and R_r is a finite set of row flat splicing rules. The picture language $L(\mathcal{A})$ generated by $\mathcal{A}$ is iteratively defined as follows:

$$f^0(M) = M; \text{For } \ i \geq 0, f^{i+1}(M) = f^i(M) \cup f(f^i(M))$$

$$L(\mathcal{A}) = f^*(M) = \bigcup_{i \geq 0} f^i(M)$$

The family of picture languages generated by array flat splicing systems is denoted by $L(AFS)$.

Example 6.2 ([27]). Consider the array flat splicing system $\mathcal{A}_R$ with alphabet $\{a, b\}$ and the initial set M consisting of the array $M_0 = \begin{smallmatrix} a & b \\ a & b \end{smallmatrix}$. The column flat splicing rule is c where $c = \left(\begin{smallmatrix} a \\ a \end{smallmatrix} \Big| \begin{smallmatrix} a \\ a \end{smallmatrix} - \begin{smallmatrix} b \\ b \end{smallmatrix} \Big| \begin{smallmatrix} b \\ b \end{smallmatrix}\right)$. The row flat splicing rules are r_1, r_2, r_3 where $r_1 = (aa|aa - aa|aa)$, $r_2 = (ab|ab - ab|ab), r_3 = (bb|bb - bb|bb)$.

The column flat splicing rule c is applicable to the pair (M_0, M_0). Note that both the components in the pair being the same initial array M_0, the requirement of equal number of rows for the application of column flat splicing rule c is satisfied. Also, the second array in the pair begins with the column $\begin{smallmatrix} a \\ a \end{smallmatrix}$ and ends with the column $\begin{smallmatrix} b \\ b \end{smallmatrix}$, as required in the rule c. The first array is "cut" between the columns $\begin{smallmatrix} a \\ a \end{smallmatrix}$ and $\begin{smallmatrix} b \\ b \end{smallmatrix}$ while the second array is "inserted" between them, yielding the array $\begin{smallmatrix} a & a & b & b \\ a & a & b & b \end{smallmatrix}$. On the other hand, the row flat splicing rules r_1, r_2 can be used to expand the array row-wise. For example, the sequence of rules r_1, r_2, r_3 could be applied to the pair of arrays with both components having the same array $\begin{smallmatrix} a & a & b & b \\ a & a & b & b \end{smallmatrix}$. Again note that both the components in the pair being the same array, the requirement of equal number of columns for the application of a sequence of row flat splicing rules is satisfied. In fact, the first array is "cut" between the first row $a\,a\,b\,b$ and the second row which is also $a\,a\,b\,b$. The second array satisfies the requirements of the sequence of rules r_1, r_2, r_3. The second array therefore can be "inserted" into the first array to yield the array

$$\begin{matrix} a\,a\,b\,b \\ a\,a\,b\,b \\ a\,a\,b\,b \\ a\,a\,b\,b \end{matrix}$$

Proceeding like this, we compute the successive terms $f^0(M)$, $f^1(M)$, ... For example,

$$f^0(M) = M = \left\{ \begin{matrix} a\,b \\ a\,b \end{matrix} \right\}, \; f^1(M) = \left\{ \begin{matrix} a\,b \\ a\,b \end{matrix}, \begin{matrix} a\,a\,b\,b \\ a\,a\,b\,b \end{matrix}, \begin{matrix} a\,b \\ a\,b \\ a\,b \\ a\,b \end{matrix}, \ldots \right\}$$

and so on. Thus, the picture language $L(\mathcal{A}_R) = f^*(M)$ consists of rectangular arrays of an even side length over the symbols a, b. Also, if the number of columns in such an array is $2n$, then the first n columns are over a while the next n columns are over b.

One such picture array is shown in Fig. 6.2.

```
a a a b b b
a a a b b b
a a a b b b
a a a b b b
```

Fig. 6.2. A member of the language of $\mathcal{A}_R$.

6.5. Pure 2D Context-Free Grammars and Languages

We next deal with the 2D grammar called pure 2D context-free grammar introduced [25] for picture array generation. A desirable feature of this model is that shearing effect in replacing a subarray of a given array is taken care of by rewriting a row or column of symbols in parallel by equal length strings and by using only terminal symbols as in a pure string grammar.

Definition 6.2 ([25]). A pure 2D context-free grammar ($P2DCFG$) is a 4-tuple $G = (\Sigma, P_c, P_r, M_I)$, where Σ is a fine set of symbols; $P_c = \{t_{c_i}/1 \leq i \leq m\}$; $P_r = \{t_{r_j}/1 \leq j \leq n\}$; Each t_{c_i},$(1 \leq i \leq m)$, called a column table, is a finite set of context-free rules of the form $a \to \alpha$, $a \in \Sigma, \alpha \in \Sigma^*$ such that for any two rules $a \to \alpha$, $b \to \beta$ in t_{c_i}, we have $|\alpha| = |\beta|$ where $|\alpha|$ denotes the length of α. Each t_{r_j},$(1 \leq j \leq n)$, called a row table, is a finite set of context-free rules of the form $c \to^t \gamma$, $c \in \Sigma, \gamma \in \Sigma^*$ such that for any two rules $c \to^t \gamma$, $d \to^t \delta$ in t_{r_j}, we have $|\gamma| = |\delta|$; $M_I \subseteq \Sigma^{**} - \{\lambda\}$ is a finite set of axiom arrays.

Derivations are defined as follows: for any two arrays M_1, M_2, we write $M_1 \Rightarrow M_2$ if M_2 is obtained from M_1 by either rewriting a column of M_1 by rules of some column table t_{c_i} in P_c or a row of M_1 by rules of some row table t_{r_j} in P_r. Here, "$\Rightarrow^*$" is the reflexive transitive closure of "$\Rightarrow$".

The picture array language $L(G)$ generated by G is the set of rectangular picture arrays $\{M/M_0 \Rightarrow^* M \in \Sigma^{**}$ for some $M_0 \in M_I\}$. The family of picture array languages generated by pure 2D context-free grammars is denoted by $P2DCFL$.

$$
M = \begin{array}{ccc} a & . & a \\ c & b & c \\ a & . & a \end{array} \Rightarrow \begin{array}{cccc} a & . & . & a \\ c & b & b & c \\ a & . & . & a \end{array} \Rightarrow \begin{array}{cccc} a & . & . & a \\ a & . & . & a \\ c & b & b & c \\ a & . & . & a \\ a & . & . & a \end{array} \Rightarrow \begin{array}{cccc} a & . & . & a \\ a & . & . & a \\ a & . & . & a \\ c & b & b & c \\ a & . & . & a \\ a & . & . & a \\ a & . & . & a \end{array} \Rightarrow \begin{array}{ccccc} a & . & . & . & a \\ a & . & . & . & a \\ a & . & . & . & a \\ c & b & b & b & c \\ a & . & . & . & a \\ a & . & . & . & a \\ a & . & . & . & a \end{array} = M_1
$$

Fig. 6.3. Derivation $M \Rightarrow^* M_1$.

Example 6.3 ([25]). Consider the pure 2D context-free grammar

$$
G = \left(\Sigma = \{a, b, c\}, P_c = \{t_{c_1}\}, P_r = \{t_{r_1}\}, M = \begin{array}{ccc} a & . & a \\ c & b & c \\ a & . & a \end{array} \right)
$$

where $t_{c_1} = \{. \to .., b \to bb\}$, and $t_{r_1} = \left\{ b \to \begin{smallmatrix} . \\ b \\ . \end{smallmatrix}, c \to \begin{smallmatrix} a \\ c \\ a \end{smallmatrix} \right\}$. A sample derivation $M \Rightarrow M_1$ on using $t_{c_1}, t_{r_1}, t_{r_1}, t_{c_1}$, in this order, is shown in Fig. 6.3. Each of the arrays occurring in the derivation given belongs to the picture language generated by G.

6.6. Flat Splicing Pure Context-Free Array Grammar Systems

Array grammar systems with 2D context-free grammar ($2CFG$) or 2D regular grammar ($2RLG$) rules in the components generating in parallel picture arrays and arrays in any two components "communicating" by array flat splicing has been considered [21], [22] and properties on generative power have been established.

We now introduce a new array grammar system called flat splicing pure context-free array grammar system by taking pure 2D context-free grammar type of rules in the components and picture array flat splicing for communication among components and exhibit its generative power by comparing with other picture generating models.

Definition 6.3. A flat splicing pure context-free array grammar system ($FSPCFAGS$) is a construct $\Gamma = (\Sigma, (P_{c_1}, P_{r_1}, M_1), \ldots (P_{c_n}, P_{r_n}, M_n), F)$, where

- Σ is a finite set of symbols.
- For each $i = 1, \ldots, n$, P_{c_i} is a finite set of column tables and P_{r_i} is a finite set of row tables with each column table and each row table consisting of context-free rules as in a $P2DCFG$.
- For $i = 1, \ldots, n$, each M_i is a finite set of rectangular axiom arrays over Σ.
- R is a finite set of column or row flat splicing rules of the form $F = (\alpha|\gamma - \delta|\beta)$, where α, β, γ and δ is either of the form ab or $\begin{smallmatrix} a \\ b \end{smallmatrix}$, for some $a, b \in \Sigma$.

A derivation in Γ takes place as follows: Each component grammar starts from an axiom array in the component and generate an array over Σ using the rules in the column/row tables in the respective component; the derivations are done with the component grammars working in parallel. At any instant, the array X generated in the ith component for some $1 \leq i \leq n$ and the array Y generated in the jth component for some $1 \leq j \leq n$ can be flat spliced using column/row flat splicing rules, thus yielding an array Z in the ith component. In fact, Z will have Y inserted in (in particular, row/column catenated with) X as directed by a finite sequence of flat splicing rules. In the other components (other than i), the arrays generated at the instant will remain unchanged during this flat spicing process. There is no priority between the rewriting step and the flat splicing step.

The picture array language $L_i(\Gamma)$ generated by the ith component of Γ consists of all rectangular arrays, generated over Σ, by derivations as described above. Without loss of generality, we choose the language of the first component as the picture array language of the system. The family of individual picture array languages generated by an $FSPCFAGS$ with at most n components is denoted by $L_n(FSPCFAGS)$.

We give an example illustrating the working of an $FSPCFAGS$ having two components.

Example 6.4. Consider an $FSPCFAGS$ having an alphabet $\{u, v, x, y, z\}$ and two components. The first component has a column table $\{x \to ux\}$ and another column table $\{x \to uz\}$ and axiom array

$$\begin{array}{c} u\,x \\ u\,x \\ u\,x \\ u\,x \end{array}$$

The second component has a column table $\{v \to vy\}$ and axiom array

$$\begin{array}{c} v\,y \\ v\,y \\ v\,y \\ v\,y \end{array}$$

The column flat splicing rule of the system is $\left(\begin{smallmatrix} u \\ u \end{smallmatrix} \middle| \begin{smallmatrix} v \\ v \end{smallmatrix} - \begin{smallmatrix} y \\ y \end{smallmatrix} \middle| \begin{smallmatrix} z \\ z \end{smallmatrix}\right)$. While in the first component an array of the form

$$\begin{array}{c} u^n\ z \\ u^n\ z \\ u^n\ z \\ u^n\ z \end{array}$$

is generated starting from the axiom array and repeatedly applying the rule in the column table $\{x \to ux\}$ in this component and finally applying the rule in the column table $\{x \to uz\}$, in the second component at the same time an array of the form

$$\begin{array}{c} v\ y^n \\ v\ y^n \\ v\ y^n \\ v\ y^n \end{array}$$

is generated starting from the axiom array and repeatedly applying the rule in the column table in this component. At this point, if the column flat splicing rule is applied, then an array of the form

$$\begin{matrix} u^n \; v \; y^n \; z \\ u^n \; v \; y^n \; z \\ u^n \; v \; y^n \; z \\ u^n \; v \; y^n \; z \end{matrix}$$

is generated. Note that arrays of the form

$$\begin{matrix} u^n \; x \\ u^n \; x \\ u^n \; x \\ u^n \; x \end{matrix}$$

generated in the first component are also included in the language generated.

The picture array language generated by the system is

$$\left\{ \begin{matrix} u^n \; x \\ u^n \; x \\ u^n \; x \\ u^n \; x \end{matrix} \middle| \, n \geq 1 \right\} \cup \left\{ \begin{matrix} u^n \; z \\ u^n \; z \\ u^n \; z \\ u^n \; z \end{matrix} \middle| \, n \geq 1 \right\} \cup \left\{ \begin{matrix} u^n \; v \; y^n \; z \\ u^n \; v \; y^n \; z \\ u^n \; v \; y^n \; z \\ u^n \; v \; y^n \; z \end{matrix} \middle| \, n \geq 1 \right\}.$$

Theorem 6.1. $L(P2DCFG) = L_1(FSPCFAGS) \subset L_2$ $(FSPCFAGS)$.

Proof. The equality $L(P2DCFG) = L_1(FSPCFAGS)$ follows from the definition of $FSPCFAGS$. In fact, the column and row tables of context-free rules and the axiom arrays of the $P2DCFG$ generating the given $P2DCFL$ are included in the only component of the $FSPCFAGS$ and the set of array flat splicing rules can be made empty.

The inclusion $L_1(FSPCFAGS) \subset L_2(FSPCFAGS)$ is clear from definitions. The proper inclusion can be seen as follows: Consider the $FSPCFAGS$ $\Gamma_1 = (\Sigma, (P_{c_1}, P_{r_1}, M_1), (P_{c_2}, P_{r_2}, M_2), F)$, where $\Sigma = \{a, b, c, d, x, z\}$. The column tables of P_{c_1} are $t_1 = \{x \to axa, a \to bab\}$, $t_2 = \{x \to z, a \to a\}$ and column table of P_{c_2} is $t_3 = \{d \to da, c \to ca\}$. The row tables of P_{r_1} are

$t_4 = \{x \to^t (axa), a \to^t (bab)\}$ and the row tables of P_{r_2} are $t_5 = \{d \to^t (cdc), a \to^t (aaa)\}$. The column flat splicing rule of F is $\{({}^t(aa)|{}^t(cc) -{}^t (aa)|{}^t(bb)), ({}^t(az)|{}^t(cd) -{}^t (aa)|{}^t(ba)), ({}^t(za)|{}^t(dc) -{}^t (ca)|{}^t(ab))\}$. The axiom array of M_1 is $\begin{smallmatrix} b & a & b \\ a & x & a \\ b & a & b \end{smallmatrix}$ and M_2 is $\begin{smallmatrix} c \\ d \\ c \end{smallmatrix}$. The derivation starts with $\begin{smallmatrix} b & a & b \\ a & x & a \\ b & a & b \end{smallmatrix}$ in the first component and with $\begin{smallmatrix} c \\ d \\ c \end{smallmatrix}$ in the second component. The rewriting in the components is done in parallel. Application of the rules of column table t_1 in the first component generate arrays of the form $\begin{smallmatrix} b^n & a & b^n \\ a^n & x & a^n \\ b^n & a & b^n \end{smallmatrix}$, where as application of the rules of the column table t_3 in the second component generate arrays of the form $\begin{smallmatrix} ca^n \\ da^n \\ ca^n \end{smallmatrix}$. Once the rule of the column table t_2 is applied, the first component arrays of the form $\begin{smallmatrix} b^n & a & b^n \\ a^n & z & a^n \\ b^n & a & b^n \end{smallmatrix}$ are generated. At this stage, column flat splicing rules of the system will be applicable, provided in both components the arrays have the same number of rows. Application of the rules of the row table t_4 in the first component and t_5 in the second component extends the respective generated arrays row wise. An array generated by Γ_1 is shown in Fig. 6.4. Note that in the first component once the column table t_2 is applied, then row table t_4 is not applicable any more and so expanding the number of rows in the arrays generated in both the components will have to take place prior to the application of the column flat splicing rule. Also, the number of rows in the arrays in both components is to be the same to enable the application of the column flat splicing operation. But this picture language $L(\Gamma_1)$ cannot be generated by any *P2DCFG* [25] as a *P2DCFG* cannot handle the context-sensitive feature in the columns of the arrays generated. This implies that

```
b b a c a a b b
b b a c a a b b
a a z d a a a a
b b a c a a b b
b b a c a a b b
```

Fig. 6.4. A picture array generated in Example 6.4.

$L(\Gamma_1)$ is not in $L_1(FSPCFAGS)$, thereby proving the proper inclusion.

We now compare with the picture language class generated by $T2CFGs$, denoting this picture language class by $T2CFL$. The subclass generated by $2CFGs$ is denoted by $2CFL$. □

Theorem 6.2. $L_2(FSPCFAGS) \setminus T2CFL \neq \emptyset$.

Proof. Consider an $FSPCFAGS$ having an alphabet $\{a, b, c, d, x, y\}$ and two components. The first component has two column tables $\{x \rightarrow axd\}$, $\{x \rightarrow azd\}$ and an axiom array $\begin{smallmatrix} a & x & d \\ a & x & d \end{smallmatrix}$. The second component has a column table $\{y \rightarrow byc\}$ and axiom array $\begin{smallmatrix} b & y & c \\ b & y & c \end{smallmatrix}$. The column flat splicing rule of the system is $\left(\begin{smallmatrix} z \\ z \end{smallmatrix}\middle|\begin{smallmatrix} b \\ b \end{smallmatrix} - \begin{smallmatrix} c \\ c \end{smallmatrix}\middle|\begin{smallmatrix} d \\ d \end{smallmatrix}\right)$. The picture array language generated by the system is $\left\{\begin{smallmatrix} a^n & x & d^n \\ a^n & x & d^n \end{smallmatrix} / n \geq 1\right\} \cup \left\{\begin{smallmatrix} a^n & z & d^n \\ a^n & z & d^n \end{smallmatrix} / n \geq 1\right\} \cup \left\{\begin{smallmatrix} a^n & z & b^n & y & c^n & d^n \\ a^n & z & b^n & y & c^n & d^n \end{smallmatrix} / n \geq 1\right\}$. But this language with a context-sensitive feature in the columns cannot be generated by any tabled $2CFG$ [24].

As a consequence of the fact $2CFL \subset T2CFL$, the following corollary can be stated. □

Corollary 6.1. $L_2(FSPCFAGS) \setminus T2CFL \neq \emptyset$.

6.7. Application

Floor designs called *kolam patterns* [2] are attractive art forms which have drawn the attention of computer science researchers interested in developing 2D picture grammars for describing images [23]. One such kolam pattern is shown in Fig. 6.5, made of primitive patterns shown in Fig. 6.6. The technique employed [23] in describing such kolam patterns is to associate suitable primitive parts of the patterns with the terminal symbols and replace these symbols in the picture arrays generated by the corresponding patterns. Utilizing this technique, here we construct a flat splicing pure context-free array grammar system G_{kolam} with two components in order to generate the picture language consisting of picture arrays that describe

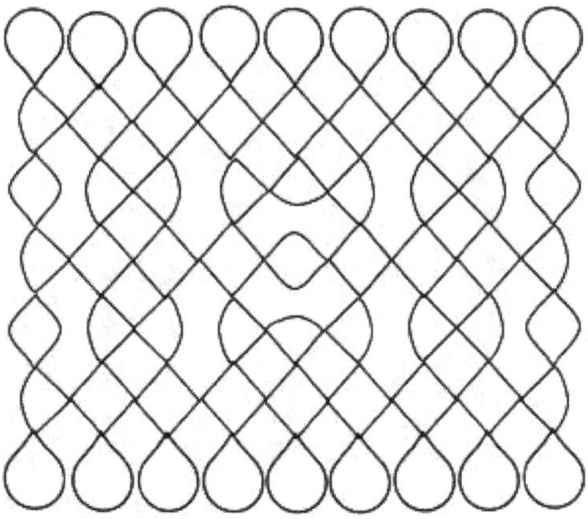

Fig. 6.5. A kolam pattern.

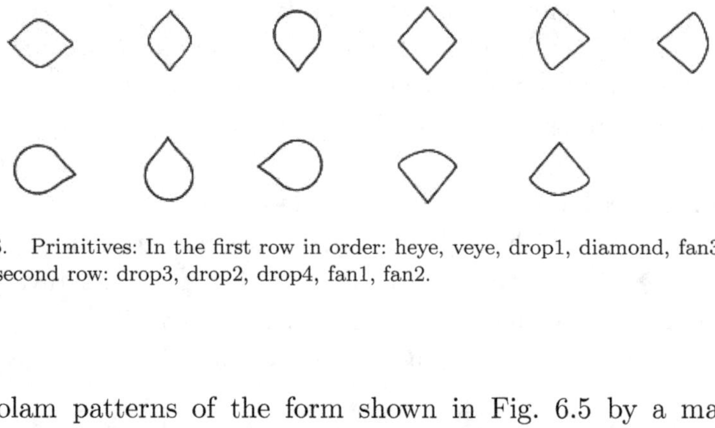

Fig. 6.6. Primitives: In the first row in order: heye, veye, drop1, diamond, fan3, fan4 ; In the second row: drop3, drop2, drop4, fan1, fan2.

the kolam patterns of the form shown in Fig. 6.5 by a mapping of primitive patterns in Fig. 6.6 to the terminal symbols of the $FSP2DCFAGS$ constructed.

The grammar system G_{kolam} is given by

$$G_{\text{kolam}} = \{\Sigma, (P_{c_1}, M_1), (P_{c_2}, P_{r_2}, M_2), F\}$$

$$\Sigma = \{a, b, c, d, e, f, g, h, i, j, k, x, y, z, c', d', g', i', j', k', j'', k''\}$$

$$P_{c_1} = \{\emptyset\}, \quad P_{c_2} = \{t_1\}$$

$$M_1 = \begin{matrix} c\ c \\ h\ e \\ b\ b \\ x\ e \\ b\ b \\ h\ e \\ g\ g \end{matrix}, \quad M_2 = \begin{matrix} c\ c'\ c \\ d\ d'\ d \\ h\ j'\ e \\ y\ i'\ y \\ h\ k'\ e \\ d\ d'\ d \\ g\ g'\ g \end{matrix}$$

$$t_1 = \{c' \to cc'c, d' \to dd'd, j' \to ej''h, i' \to yi'y,$$
$$k' \to ek''h, g' \to gc'g, j'' \to hj'e, k'' \to hk'e,\}$$
$$R = \{((ch)^t|(cd)^-(cd)^t|(ce)^t), ((hb)^t|(dh)^t - (de)^t|(eb)^t),$$
$$((bx)^t|(hy)^t - (ey)^t|(be)^t), ((xb)^t|(yh)^t - (ye)^t|(eb)^t),$$
$$((bh)^t|(hd)^t - (ed)^t|(be)^t), ((hg)^t|(dg)^t - (dg)^t|(eg)^t)\}$$

In the first component, the generation process starts with the given axiom M_1, while in the second component with M_2. Subsequent steps of the generation or derivation take place in the two components together in parallel, with the rules of table t_1 being applied. The first component contains arrays of the form shown in Fig. 6.7(i), while in the second component, the rules of table t_1 generate arrays, one member of which is shown in Fig. 6.7(ii). At this stage, the flat splicing rules become applicable and yield picture arrays in the first component, one member of which is shown in Fig. 6.7(iii). If we define a mapping ϕ that assigns a primitive pattern for each terminal

$$\begin{matrix} c\ c \\ h\ e \\ b\ b \\ x\ e \\ b\ b \\ h\ e \\ g\ g \end{matrix} \qquad \begin{matrix} c\ c\ c\ c\ c\ \ c\ \ c \\ d\ d\ d\ c\ d\ \ d\ \ d \\ h\ e\ h\ j\ e\ \ h\ \ e \\ y\ y\ y\ i\ y\ \ y\ \ y \\ h\ e\ h\ k\ e\ \ h\ \ e \\ d\ d\ d\ d\ d\ \ d\ \ d \\ g\ g\ g\ g\ g\ \ g\ \ g \end{matrix} \qquad \begin{matrix} c\ c\ c\ c\ \ c\ \ c\ c\ c\ c \\ h\ d\ d\ d\ \ d\ \ d\ d\ d\ e \\ b\ h\ e\ h\ \ j\ \ e\ h\ e\ b \\ x\ y\ y\ y\ \ i\ \ y\ y\ y\ e \\ b\ h\ e\ h\ \ k\ \ e\ h\ e\ b \\ h\ d\ d\ d\ \ d\ \ d\ d\ d\ e \\ g\ g\ g\ g\ \ g\ \ g\ g\ g\ g \end{matrix}$$

(i) (ii) (iii)

Fig. 6.7. (i) A picture array generated in the first component of G_{kolam}, prior to flat splicing. (ii) A picture array generated in the second component of G_{kolam}, prior to flat splicing. (iii) A picture array generated in the first component of G_{kolam} on flat splicing.

Fig. 6.8. Left: Pattern generated in the first component. Right: Pattern generated in the second component.

symbol, then the mapping can be extended to the picture array by applying this to each entry in the array. This will then yield the desired pattern. Here we define $\phi(d) = \phi(d') = diamond$, $\phi(i) = \phi(i') = heye$, $\phi(j) = \phi(j') = \phi(j'') = fan1$, $\phi(k) = \phi(k') = \phi(k'') = fan2$, $\phi(c) = \phi(c') = drop1$, $\phi(g) = \phi(g') = drop2$, $\phi(a) = drop3$, $\phi(f) = drop4$, $\phi(b) = veye$ $\phi(h) = fan3$, $\phi(e) = fan4$.

Then an application of ϕ to the picture array in Fig. 6.7(iii) will yield the kolam pattern in Fig. 6.5. Note that the kolam patterns of the picture arrays prior to flat splicing are also shown in Fig. 6.8.

Acknowledgements

G. Samdanielthompson acknowledges with gratitude the award (No.: F1-17.1/2016-17/MANF-2015-17-TAM-51257/(SA-III/Website)) of Maulana Azad National Fellowship for minority students by UGC, India which has enabled him to carry out his research for PhD in the Department of Mathematics, Madras Christian College under the supervision of Dr. N. Gnanamalar David. K. G. Subramanian is grateful to UGC, India for the award of Emeritus Fellowship (No.F.6-6/2016-17/EMERITUS-2015-17-GEN-5933/(SA-II)) to him to execute his work in the Department of Mathematics, Madras Christian College.

Bibliography

[1] Adleman, L. M. Molecular computation of solutions to combinatorial problems. *Science* **266**(5187), 1021–1024, 1994.

[2] Ascher, M. The Kolam Tradition: A tradition of figure-drawing in southern India expresses mathematical ideas and has attracted the attention of computer science. *Am. Sci.* **90**(1), 56–63, 2002.

[3] Bersani, M. M., Frigeri, A. and Cherubini, A. Expressiveness and complexity of regular pure two-dimensional context-free languages. *Int. J. Comput. Math.* **90**(8), 1708–1733, 2013.

[4] Berstel, J., Boasson, L. and Fagnot, I. Splicing systems and the Chomsky hierarchy. *Theoret. Comput. Sci.* **436**, 2–22, 2012.

[5] Chandra, P. H., Subramanian, K. G. and Thomas, D. G. Parallel splicing on images. *Int. J. Pattern Recogn. Artificial Intell.* **18**(06), 1071–1091, 2004.

[6] Csuhaj-Varjú, E., Dassow, J., Kelemen, J. and Păun, Gh. (eds). *A Grammatical Approach to Distribution and Cooperation.* Gordon and Breach Science Publishers, 1994.

[7] Dassow, J. and Mitrana, V. Evolutionary grammars: a grammatical model for genome evolution. In *Bioinformatics.* Springer, Berlin, pp. 199–209, 1996.

[8] Esik, Z., Martin-Vide, C. and Mitrana, V. (eds). *Recent Advances in Formal Languages and Applications.* Studies in Computational Intelligence. Springer-Verlag, 2006.

[9] Freund, R., Kari, L. and Păun, Gh. DNA computing based on splicing: The existence of universal computers. *Theoret. Comput. Syst.* **32**(1), 69–112, 1999.

[10] Giammarresi, D. and Restivo, A. Two-dimensional languages. In *Handbook of Formal Languages.* Springer, Berlin, pp. 215–267, 1997.

[11] Head, T. Formal language theory and DNA: an analysis of the generative capacity of specific recombinant behaviors. *Bull. Math. Biol.* **49**(6), 737–759, 1987.

[12] Head, T., Păun, Gh. and Pixton, D. Language theory and molecular genetics: generative mechanisms suggested by DNA recombination. In *Handbook of Formal Languages.* Springer, Berlin, pp. 295–360, 1997.

[13] Krithivasan, K., Chakaravarthy, V. T. and Rama, R. Array splicing systems. In *New Trends in Formal Languages.* Springer, Berlin, pp. 346–365, 1997.

[14] Martin-Vide, C., Mitrana, V. and Păun, Gh. (eds). *Formal Languages and Applications*, Springer, 2004.

[15] Maurer, H. A., Salomaa, A. and Wood, D. Pure grammars. *Inform. Control* **44**(1), 47–72, 1980.

[16] Păun, Gh. Computing with membranes. *J. Comput. Syst. Sci.* **61**(1), 108–143, 2000.

[17] Păun, Gh., Rozenberg, G. and Salomaa, A. *DNA Computing — New Computing Paradigms.* Texts in Theoretical Computer Science: An EATCS Series. Springer, 1998.

[18] Rosenfeld, A. *Picture Languages.* Academic Press, Reading, MA, 1979.

[19] Rosenfeld, A. and Siromoney, R. Picture languages — a survey. *Lang. Des.* **1**(3), 229–245, 1993.

[20] Rozenberg, G. and Salomaa, A. (eds). *Handbook of Formal Languages*, Vols. 1–3. Springer, Berlin, 1997.

[21] Subramanian, K. G., Samdanielthompson, G., David, N. G. and Nagar, A. K. Generating picture arrays based on grammar systems with flat splicing operation. In *Innovations in Bio-inspired Computing and Applications*. Advances in Intelligent Systems and Computing, Vol. 424. Springer International Publishing, Switzerland, pp. 251–261, 2016.

[22] Samdanielthompson, G., David, N. G. and Subramanian, K. G. Flat splicing context free array grammar systems generating picture arrays. *Int. J. Artificial Intell. Soft Comput.* **5**(4), 294–310, 2016.

[23] Siromoney, G., Siromoney, R. and Krithivasan, K. Abstract families of matrices and picture languages. *Comput. Graph. Imag. Process* **1**(3), 284–307, 1972.

[24] Siromoney, R., Subramanian, K. G. and Rangarajan, K. Parallel/sequential rectangular arrays with tables. *Int. J. Comput. Math.* **6**(2), 143–158, 1977.

[25] Subramanian, K. G., Ali, R. M., Geethalakshmi, M. and Nagar, A. K. Pure 2D picture grammars and languages. *Discrete Appl. Math.* **157**(16), 3401–3411, 2009.

[26] Subramanian, K. G., Roslin Sagaya Mary, A. and Dersanambika, K. S. Splicing array grammar systems. In *International Colloquium on Theoretical Aspects of Computing*. Springer, Berlin, 2005.

[27] Subramanian, K. G., Nagar, A. K. and Pan, L. A picture array generating model based on flat splicing operation. In *Proceedings of the 10th International Conference on Bio-inspired Computing: Theories and Applications (BICTA)*, 2015.

Chapter 7

Fingerprint Minutiae to Fixed-Length Bit-String: A Review of Recent Development

Zhe Jin*,§, Wei-Jing Wong† and Andrew Beng Jin Teoh‡,¶

*School of Information Technology, Monash University Malaysia
Selangor Darul Ehsan, Malaysia

†Swinburne University of Technology (Sarawak Campus), Malaysia

‡Yonsei University, South Korea

§jinzhe@monash.edu

¶bjteoh@yonsei.ac.kr

Some biometric cryptographic applications such as fuzzy identity-based identification (FIBI) and fuzzy commitment require ordered and fixed-length bit-string-like IrisCode as input. However, fingerprint minutiae representation (e.g. ISO minutiae format) is unordered and variable in size. Such a characteristic is inapplicable to the aforementioned applications. One of the feasible solutions is to convert minutiae into ordered and fixed-length bit-string, namely *point-to-string conversion*. The point-to-string conversion has attracted much attention and a number of proposals have been reported in literature over the past decade. Furthermore, the topic of point-to-string conversion continues to gain the interest from the research community lately. In this chapter, the point-to-string conversion methods proposed in early stage are revisited to be served as a background study. Thereafter, a review of recent development on point-to-string conversion is presented. More specifically, two recently proposed methods (i.e. Kernel-Learning and Bag-of-Minutiae) are introduced in detail. Finally, conclusion is given to summarize the challenges, and future prospect in this research topic.

7.1. Introduction

Fingerprint is one of the most popular traits in deployed biometric systems for identity verification and identification [9]. The features extracted from fingerprint can be divided into three levels: namely *level* 1 (i.e. Global level), *level* 2 (i.e. Local level), and *level* 3 (i.e. Very-fine level). The features at the global level are based on the pattern of fingerprint ridge line flow, typically *singular points* (e.g. loop and delta) and *coarse ridge shape* (i.e. Arch, Loop and Whorl). Global level features are useful for fingerprint classification and indexing but less discriminative for accurate matching [22]. Features at the local level are local ridge characteristics, namely *minute details.* There are two most prominent ridge characteristics, called *ridge bifurcations* and *ridge ending* as shown in Fig. 7.1(a). Generally, fingerprint minutiae are stable and robust to fingerprint impression conditions [22]. Features at very-fine level refer to the intraridge details including width, shape, curvature, edge contours of ridges, etc. The most significant very-fine level feature is the finger *sweat pores.* However, very-fine level features can only be extracted from a high-resolution (e.g. 1000 dpi) fingerprint image with good quality and unavailable to the most practical applications.

Each minutia can be associated with a number of attributes, including location coordinates, orientation, type (e.g. ridge ending or ridge bifurcation), a weight based on the quality of the fingerprint image in the neighbourhood of each minutia, and so on. However, from the common practices, only two attributes are used to represent a minutia: x- and y-coordinates pertaining to the location of minutia in the fingerprint; the orientation θ of the ridge line to which the minutia is attached [17] as shown in Fig. 7.1(b). Besides that, fingerprint minutiae template is an unordered and variable-sized point set data due to the presence of spurious and missing genuine minutiae caused by largely rotation, translation and skin elastic deformation of fingerprint image acquired from the different impressions of the identical finger. The unordered and variable size representation restricts the biometric cryptographic applications (e.g. fuzzy identity-based identification (FIBI) [35] and fuzzy commitment [16]) that only admit ordered and fixed-length binary vector form. This restriction motivates a new research topic, namely *point-to-string conversion*, which is devoted to devise feasible solutions for transforming unordered and variable size data into ordered and fixed-length feature vector. Although, a variety of techniques have been proposed to address the point-to-string conversion problem, there remains much room for further development.

The organization of this chapter is as follows: a briefing of point-to-string conversion methods on early stage is given in Section 7.2. Then, point-to-string conversion methods based on the kernel learning method and bag-of-minutiae are described in Sections 7.3 and 7.4, respectively. Finally, conclusion is provided in Section 7.5.

7.2. Briefs on the Point-to-String Conversion Methods on Early Stage

In literature, point-to-string conversion techniques can be divided into three categories [11]: (1) reference-based approach; (2) histogram-based approach; and (3) spectral transform approach.

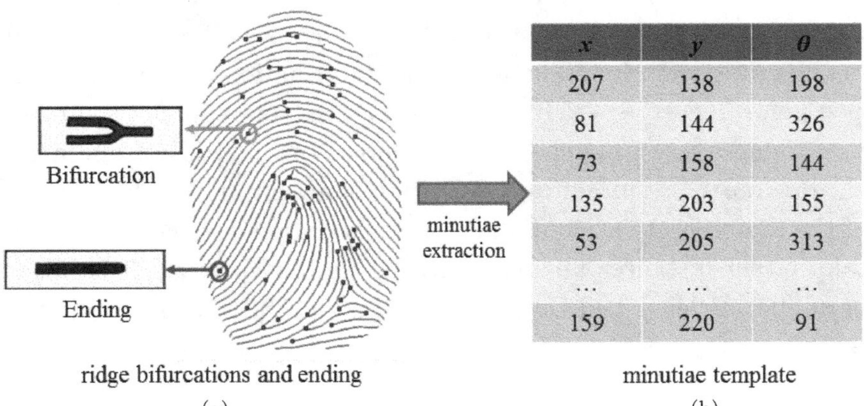

Fig. 7.1. (a) Illustrate ridge bifurcations and ridge ending; (b) minutiae template including (x, y) coordinates and the orientation, θ.

Reference-based approach defines a fixed-size reference, e.g. circumference of a circle [32, 33], N random cuboids [34]. Then the biometric features are quantized into fixed-length representation according to the specific reference. For the histogram-based approach, the fixed-length representation is generated based on the histogram, which is formed based on the tabulated frequencies of extracted features erected over discrete intervals on the extracted feature space. In the spectral transform approach, specific spectral transform technique, such as Fourier transform, is used to transform minutiae to corresponding domain, so that the fixed-length representation can be generated using specific analytical methods in that domain.

7.2.1. *Reference-based approach*

Sutcu *et al.* [32, 33] proposed a geometric transformation to convert fingerprint minutiae into a fixed-length feature vector. This method uses the circumference of a circle as a reference and divides it into m equal-width arcs. For every minutia pair, a straight line passing through these two points and mark its intersections onto the circumference of the circle. An m-dimensional integer feature vector is then constructed by counting the number of projected minutiae in the respective arcs. One limitation of this method is that the transformation is not rotation-invariant, thus the fingerprints have to be aligned before transformation. The additional information such as registration point (e.g. core or delta point) is required for aligning two fingerprint images to be matched. Furthermore, the analysis of minimum entropy on the resultant representation is absent. Thus, the security of the representation under different attack scenarios remains uncertain. Other instances in this approach, e.g. Sutcu *et al.* [34], Nagar *et al.* [23], Bringer and Despiegel [4], Liu *et al.* [21], can be found in literature.

7.2.2. *Histogram-based approach*

Farooq *et al.* [6] generated a binary fingerprint representation based on the histograms of triangular features generated from minutiae triplets. Seven invariant features: length of three sides, three angles between each side and each minutia orientation and height of the triangle are extracted and quantized into 24 bits, which yields a 2^{24}-bit binary string. However, this method incurs a high computational cost due to the exhaustive calculation of features for all possible minutiae triplets. Despite anonymous and revocable representations is the main initiative for Farooq's approach, it can be adopted as a useful technique for fingerprint minutiae template protection [13, 37–39].

7.2.3. *Spectral transform approach*

Xu *et al.* [41] proposed a Spectral Minutiae approach to convert a set of minutiae into a fixed-length feature vector. The proposed approach performs Fourier transform on a minutia set and remaps the Fourier spectral magnitude onto a polar-logarithmic coordinate. By doing so, the spectral minutiae representation is invariant to rotation, shifting and scaling variations. An analytical representation for minutiae is further proposed to minimize error, which can directly be evaluated on polar-logarithmic grids. As the number of grids is fixed, a fixed-length representation can be derived. However, the accuracy of this approach over point-to-point (minutiae) and two-stage procedure matching (minutia descriptor) approaches is inferior. Nandakumar [24] uses a phase spectrum, instead of magnitude of minutiae, to generate fixed-length feature vector and achieves a state-of-the-art accuracy performance. However, a proper alignment is mandatory as phase spectrum is not rotation-, shifting- and scaling-invariant.

7.3. Point-to-String Conversion Based on the Kernel Learning Method

7.3.1. *Preliminary*

7.3.1.1. *Kernel principal component analysis*

Principal component analysis (PCA) is an orthogonal transformation that converts a set of correlated data into a set of linearly uncorrelated data, called principal components [29]. However, PCA assumes that the data are multivariate Gaussian-distributed. As this assumption may not be valid in practice, kernel PCA [30] has been developed. The basic idea of kernel PCA is to apply a nonlinear function $\Phi(\cdot)$ to a set of data $\mathbf{x}_j$ on the kernel space, $\mathbf{x}_j \in \mathbb{R}$, $j = 1, \ldots, l$. Let the data $\mathbf{x}_j$ to be mapped onto a feature space: $\Phi(\mathbf{x}_1), \ldots, \Phi(\mathbf{x}_l)$, where $\sum_{j=1}^{l} \Phi(\mathbf{x}_j) = 0$. The covariance matrix is computed by

$$\boldsymbol{C} = \frac{1}{l} \sum_{j=1}^{l} \Phi(\mathbf{x}_j) \Phi(\mathbf{x}_j)^{\mathrm{T}} \tag{7.1}$$

The covariance matrix is used to determine eigenvalues $\boldsymbol{\lambda}$ and eigenvectors $\mathbf{v}$ in the feature space, such that

$$\boldsymbol{\lambda}\mathbf{v} = \boldsymbol{C}\mathbf{v} \tag{7.2}$$

By substituting Eq. (7.1) into Eq. (7.2), the eigenvectors can be expressed in terms of coefficients $a_1, \ldots, a_l$ as follows:

$$\mathbf{v} = \sum_{i=1}^{l} a_i \Phi(\mathbf{x}_i) \tag{7.3}$$

By defining an $l \times l$ kernel matrix $\mathbf{K}$, such that

$$\mathbf{K} = k(x, y) = \langle \Phi(x) \cdot \Phi(y) \rangle = \Phi(x)^T \Phi(y) \tag{7.4}$$

Substituting Eq. (7.1), Eq. (7.3) and Eq. (7.4) into Eq. (7.2) yields

$$l\boldsymbol{\lambda}\boldsymbol{\alpha} = \mathbf{K}\boldsymbol{\alpha} \tag{7.5}$$

where $\boldsymbol{\alpha}$ and $\boldsymbol{\lambda}$ are the eigenvectors and eigenvalues of $\mathbf{K}$, respectively. Then, the eigenvectors are normalized using $(\boldsymbol{a}_i)^T \boldsymbol{a}_i = 1$.

The testing feature $\mathbf{x}_{\text{test}}$ is projected onto the principle component as shown in Eq. (7.6) by substituting Eq. (7.3), Eq. (7.4), where

$\mathbf{v}_{pc} \subset \mathbf{v}$ and $\boldsymbol{\alpha}_{pc} \subset \boldsymbol{\alpha}$. Examples of Mercer's admissible kernels include Linear Kernel $k(x,y) = x^T y + c$, Polynomial Kernel $k(x,y) = (ax^T y + c)^d$, and Gaussian Kernel $k(x,y) = \exp(-\|x-y\|^2/2\sigma^2)$.

$$\mathbf{Y} = \mathbf{v}_{pc}{}^T \Phi(\mathbf{x}_{\text{test}}) = \sum_{i=1}^{l} \boldsymbol{\alpha}_{pc} \Phi\left(\mathbf{x}_i\right)^T \Phi\left(\mathbf{x}_{\text{test}}\right) = \sum_{i=1}^{l} \boldsymbol{\alpha}_{pc} k(\mathbf{x}_i, \mathbf{x}_{\text{test}}) \tag{7.6}$$

7.3.1.2. *Kernelized locality-sensitive hashing*

Locality-sensitive hashing (LSH) [5, 7] is a method that searches for the approximate nearest neighbours that preserve the property: $\Pr[h(x_i) = h(x_j)] = sim(x_i, x_j)$, where h is a hash function and $sim(x_i, x_j) \in [0,1]$ is a similarity function. Charikar *et al.* [5] proposed a hash function $h_c(x)$ based on inner product similarity: $sim(x_i, x_j) = x_i^T x_j$, which converts the output of a product to a binary value using a random hyperplane:

$$h_c(x) = \begin{cases} 1 & w_c^T x > 0 \\ 0 & \text{otherwise} \end{cases} \tag{7.7}$$

where $w_c = [w_{c1}, w_{c2}, \ldots, w_{cd}]^{\mathrm{T}}$ is a random hyperplane from a zero-mean multivariate Gaussian $N(0,1)$ of dimension d. It is noted that, based on the random projection, a good performance of LSH can be obtained according to the Johnson–Lindenstrauss lemma [15].

However, when the feature space embedding is unknown or computationally infeasible, it is impossible to apply the original LSH in this case. Kulis and Grauman [18] proposed a kernelized locality-sensitive hashing (KLSH) method based on the Central Limit theorem [26]. This method allows the approximation of a random vector using a set of training data. KLSH can be summarized in the following four steps:

- Select p data points and construct a kernel matrix $\mathbf{K}$ for this data.
- Centralize the kernel matrix.

- Construct a hash table for the database items: For each hash function $h(\phi(x))$, select n indices at random from $[1, \ldots, p]$ and initiate a binary vector $\boldsymbol{e}_s$ whose ones indicate the indices, compute a weighted matrix $w = \mathbf{K}^{-1/2}\boldsymbol{e}_s$, and assign the bits according to $h(\phi(x)) = \text{sign}(\sum_i w(i)k(x, x_i))$.
- For each query, form a hash key using these hash functions and employ existing LSH methods to find the approximate nearest neighbours.

7.3.1.3. *Polar grid-based 3-tuple quantization*

Polar grid-based 3-tuple quantization (PGTQ) [14] is an alignment-free minutiae descriptor that utilizes variable-sized tessellated quantization in the polar coordinate. In this method, sectors near the reference minutia have smaller area and vice versa. This leads to a smaller (respectively, larger) quantization step around (respectively, further away from) the reference minutia to tolerate the fingerprint elastic deformation. In the original PGTQ descriptor, the polar coordinate covers the entire image and produces a lengthy bit string, which is undesirable for kernel transformation due to a high computational cost. As a solution, the polar coordinate that only covers a part of the image limited by a circle with radius R is considered. With this, the size of the resultant bit-string can be significantly reduced. The details of the modified PGTQ descriptor are described as follows:

(1) Let $m_r = \{x_r, y_r, \theta_r\}$ be the reference minutiae. The neighboring minutiae within a circle with radius R in Euclidean distance is rotated and translated based on the reference minutiae using Eqs. (7.8) and (7.9). The transformed minutiae are represented as $m^t = \{x_i^t, y_i^t, \theta_i^t | i = 1, \ldots N_R - 1\}$, where N_R is the total number of minutiae within a pre-defined radius R.

$$\begin{bmatrix} x_i^t \\ y_i^t \end{bmatrix} = \begin{bmatrix} \cos\theta_r & -\sin\theta_r \\ \sin\theta_r & \cos\theta_r \end{bmatrix} \begin{bmatrix} x_i - x_r \\ -(y_i - y_r) \end{bmatrix} \tag{7.8}$$

$$\theta_i^t = \begin{cases} \theta_i - \theta_r; & \theta_i \geq \theta_r \\ 360 + \theta_i - \theta_r; & \theta_i < \theta_r \end{cases} \tag{7.9}$$

(2) The translated and rotated minutiae are then converted into polar coordinates using Eqs. (7.10) and (7.11). Here, ρ_i and α_i indicate the radial distance (in pixels) and the radial angle of the ith minutia in Polar coordinates ($\alpha_i \in (0, 360]$), respectively.

$$\rho_i = \sqrt{(x_i^t)^2 + (y_i^t)^2} \tag{7.10}$$

$$\alpha_i = \arctan\left(\frac{y_i^t}{x_i^t}\right) \tag{7.11}$$

(3) *3-Tuple-based quantization.* The 3-tuple-based quantization is a sector-based quantization involving all minutiae in the neighbourhood. Each quantized minutia can be represented as a vector $\omega = \{\rho^q, \alpha^q, \theta^q\}$, such that

$$\rho_i^q = \lfloor \rho_i / x \rfloor \tag{7.12}$$

$$\alpha_i^q = \lfloor \alpha_i / y \rfloor \tag{7.13}$$

$$\theta_i^q = \lfloor \theta_i / z \rfloor \tag{7.14}$$

where "/" denotes quotient; x, y and z indicate the radius for each polar grid (in pixels), radial angle for tolerance ($y \in (0, 360]$) and orientation angle to be tolerated $z \in (0, 360]$, respectively. The quantization level is determined by x, y and z.

(4) *Binarization.* The quantized minutiae ω are then binarized using the polar grids. A simple rule [19] is adopted. That is to map a polar grid to bit "1" if the polar grid contains more than one minutia, and otherwise, a polar grid is mapped to "0". By concatenating the individual output bits from the polar grids, a binary vector with length equal to the number of polar grids $l = \lceil 360/x \rceil \times \lceil 360/y \rceil \times \lceil 360/z \rceil$, where $\lceil \cdot \rceil$ denotes the ceiling function, is eventually obtained. The above steps are repeated by

changing the reference minutia with every remaining minutia to generate the full binary PGTQ descriptor. As the total minutiae number (N_m) extracted from each fingerprint image could be different, this template, denoted by $\boldsymbol{\Omega} \in \{0,1\}^{N_m \times l}$, is variable in size.

(5) *Matching.* To evaluate the similarity between two sets of modified PGTQ descriptors, a two-stage matching strategy that is composed of local and global matching is used. The local descriptor matching searches for the intersections between two binary strings in which the PGTQ descriptor is represented. On the other hand, the global matching is to find the ratio of the matched descriptor pairs over all potential pairs as the final matching score.

Let $\boldsymbol{\Omega}^e = [\boldsymbol{b}_1^e; \boldsymbol{b}_2^e, \ldots, \boldsymbol{b}_{n^e}^e]$ and $\boldsymbol{\Omega}^q = [\boldsymbol{b}_1^q; \boldsymbol{b}_2^q, \ldots, \boldsymbol{b}_{n^q}^q]$ be the enrolled and query descriptor sets that consist of n^e and n^q *l-bit* binary strings, respectively. From this point onwards, the notation of $\boldsymbol{b}$ is slightly abused, where $\boldsymbol{b}_{i,k}$ represents the kth bit for the ith binary string with $1 \leq k \leq l$ and $1 \leq i \leq n^e$ or n^q. Taking into account the difference of minutiae quantity in the enrolled and query image, the similarity scores between two local descriptors $\boldsymbol{\Omega}^e$ and $\boldsymbol{\Omega}^q$ are normalized as follows:

$$S_{ij}^b = \frac{(N_j^q + N_i^e)\sum_{k=1}^{l}(\boldsymbol{b}_{j,k}^q \cdot \boldsymbol{b}_{i,k}^e)}{(N_j^q)^2 + (N_i^e)^2} \tag{7.15}$$

where S^b denotes the matching score between two binary strings, $\cdot$ represents a bitwise AND operator, $N_i^e = \sum_{k=1}^{l}(\boldsymbol{b}_{i,k}^e)$ and $N_j^q = \sum_{k=1}^{l}(\boldsymbol{b}_{j,k}^q)$ denote the total number of 1's of the enrolled and query bit-strings, respectively. The term $\sum_{k=1}^{l}(\boldsymbol{b}_{j,k}^q \cdot \boldsymbol{b}_{i,k}^e)$ in Eq. (7.15) counts the bit positions that have value "1" in both query and enrolled bit-strings. The scores in matrix $S^b \in \mathbb{R}^{n^q \times n^e}$ range from 0 to 1 where "1" indicates a perfect match.

Once the similarity score matrix S^b is calculated from the local descriptor matching, a global matching process is carried out. Given the score matrix $S^b = \{s_{ij}^b\}$, the final score can be calculated as

$$S_{\mathrm{PGTQ}} = \max\left\{\frac{1}{m}\sum_j s_{j(\max)}, \frac{1}{n}\sum_i s_{i(\max)}\right\} \tag{7.16}$$

where $s_{j(\max)} = \max_i\{s_{ij}^b\}$ and $s_{i(\max)} = \max_j\{s_{ij}^b\}$ represent the maximum score component of the ith column and jth row, respectively.

7.3.2. *Kernel method-based fixed-length transformation*

Generally, the kernel learning-based point-to-string conversion consists of four main components: minutiae descriptor extraction, kernel learning-based transformation, feature binarization and matching. The fixed-length representation is induced by the fixed number of training samples and the projection matrix is produced by the kernels. The procedure of our kernel-based point-to-string conversion is described as follows:

(1) *Kernel matrix computation.* Let $\boldsymbol{\Omega} = \{\boldsymbol{\Omega}^{\mathrm{train}}(i)|i = 1, \ldots, N_{\mathrm{train}}\}$ be a set of training samples for PGTQ descriptor and N_{train} denote the total number of $\boldsymbol{\Omega}^{\mathrm{train}}$. A kernel matrix $\boldsymbol{K} \in \mathbb{R}^{N_{\mathrm{train}} \times N_{\mathrm{train}}}$ can be constructed by using the following kernel function:

$$\begin{aligned}\boldsymbol{K}(i,j) &= k\left(\boldsymbol{\Omega}^{\mathrm{train}}(i), \boldsymbol{\Omega}^{\mathrm{train}}(j)\right)\\ &= \exp(-(1 - S_{\mathrm{PGTQ}}(i,j))^2/2\sigma^2)\end{aligned} \tag{7.17}$$

where $S_{\mathrm{PGTQ}} \in [0, 1]$ is the dissimilarity measure described in Eqs. (7.15) and (7.16). Unlike Mercer's kernel matrix that admits metrics such as inner product or Euclidean distance, the kernel matrix computed with sole $S_{\mathrm{PGTQ}}(i,j)$ is indeed non-SPD, in the sense that it is neither symmetric nor positive definite.

This may potentially deteriorate the accuracy performance of kernel-based methods [25, 40]. In order to rectify this problem, following Jayasumana *et al.* [10], a heat kernel function shown in Eq. (7.17) to induce a SPD kernel function subject to a specific range of spread factor, σ is applied. Jin *et al.* [11] has shown the feasible range of σ that fulfils the SPD property and that the SPD property is an essential factor for accurate performance.

(2) With $\boldsymbol{K}$, the projection matrix $\bar{\boldsymbol{\alpha}} \in \mathbb{R}^{N_{\text{train}} \times N_{\text{dim}}}$ (i.e. the eigenvectors of KPCA or the weighted matrix of KLSH) can be produced, where N_{dim} denotes the number of desired output dimensions.

(3) *Transformation of PGTQ descriptor to fixed-length real-valued representation.* Let $\boldsymbol{\Omega}^{\text{test}} = \{\boldsymbol{\Omega}^{\text{test}}(i) | i = 1, \ldots, N_{\text{test}}\}$ be the set testing samples of the PGTQ descriptor.

Step 1: $\boldsymbol{\Omega}^{\text{test}}$ is first matched with all training samples $\boldsymbol{\Omega}^{\text{train}}(i)$ for $1 \leq i \leq N_{\text{train}}$. Subsequently, a fixed vector $\mathbf{v}^{fl} \in \mathbb{R}^{1 \times N_{\text{train}}}$ is formed by concatenating N_{train} matching scores.

$$\mathbf{v}_i^{fl} \leftarrow \text{sim}(\boldsymbol{\Omega}^{\text{test}}, \boldsymbol{\Omega}^{\text{train}}(i)) \tag{7.18}$$

Step 2: $\mathbf{v}_i^{fl}$ is then transformed into the kernel space using the kernel function defined in Eq. (7.17), in which the transformed feature vector $\bar{\mathbf{v}}^{fl}$ can be described by

$$\bar{\mathbf{v}}^{fl} = \exp(-(1 - \mathbf{v}^{fl})^2 / 2\sigma^2) \tag{7.19}$$

Step 3: The fixed-length ordered real-valued vector can be generated by projecting $\bar{\mathbf{v}}^{fl}$ using the projection matrix $\bar{\boldsymbol{\alpha}}$.

$$T^{fl} = \bar{\mathbf{v}}^{fl} \bar{\boldsymbol{\alpha}} \tag{7.20}$$

where $\bar{\mathbf{v}}^{fl} \in \mathbb{R}^{1 \times N_{\text{train}}}$ and $\bar{\boldsymbol{\alpha}} \in \mathbb{R}^{N_{\text{train}} \times N_{\text{dim}}}$. Thus, the length of T^{fl} is N_{dim}. The pseudo-code of the kernel-based transformation is presented in Algorithm 7.1.

Algorithm 7.1. Kernel-based fixed-length transformation

Input $\boldsymbol{\Omega}^{\text{train}}$, $\boldsymbol{\Omega}^{\text{test}}$, N_{dim} and σ

Stage 1: Kernel Matrix Computation

1.1: $N_{\text{train}} = \text{length}(\boldsymbol{\Omega}^{\text{train}})$

1.2: For $i = 1$: N_{train}

    Randomly toggling bits for each row of $\boldsymbol{\Omega}^{\text{train}}(\text{i})$

  End for

1.3: Initialize $\boldsymbol{K}(i,j) = 1,\ 1 \leq i,j \leq N_{\text{train}}$

1.4: For $i = 1$: N_{train}

    For $j = 1$: i

      $\boldsymbol{K}(i,j) = \text{sim}(\boldsymbol{\Omega}^{\text{train}}(\text{i}), \boldsymbol{\Omega}^{\text{train}}(\text{j}))$

      $\boldsymbol{K}(i,j) = \exp(-(1-k(i,j))^2/2\sigma^2)$

    End for

  End for

1.5: Make $\boldsymbol{K}$ symmetric

Stage 2: Compute The Projection Matrix $\bar{\alpha}$

2.1(a): $\bar{\alpha} \leftarrow \text{KPCA}(\boldsymbol{K})$ or

2.1(b): $\bar{\alpha} \leftarrow \text{KLSH}(\boldsymbol{K})$, $\bar{\alpha} \in \mathbb{R}^{N_{\text{train}} \times N_{\text{dim}}}$

Stage 3: Transform variable-size PGTQ to fixed-length vector

3.1: Initialize $\mathbf{v}^{fl}(i)$, $\bar{\mathbf{v}}^{fl}(i) = 1$; $1 \leq i \leq N_{\text{train}}$

    For $i = 1$: N_{train}

      $\mathbf{v}^{fl}(i) \leftarrow \text{sim}(\boldsymbol{\Omega}^{\text{test}}\boldsymbol{\Omega}^{\text{train}}(i))$

      $\bar{\mathbf{v}}^{fl}(i) = \exp(-\left(1-\mathbf{v}^{fl}(i)\right)^2/2\sigma^2)$

    End for

3.2: $\boldsymbol{T}^{fl} = \bar{\mathbf{v}}^{fl}\bar{\alpha}$

3.3: Save $\boldsymbol{T}^{fl}$

Output the real-valued fixed-length representation $\boldsymbol{T}^{fl}$

7.3.3. *Feature vector binarization*

Biometric feature binarization is the process of converting real-valued biometric features into a binary string. Generally, biometric binarization can be decomposed into two essential components: biometric quantization and feature encoding. These components may be governed by a static or a dynamic bit allocation algorithm determining whether the quantity of binary bits allocated to every feature dimension is fixed or varied, respectively. In this chapter, two different binarization methods, zero-thresholding quantization and reliability-based dynamic quantization method [20], are investigated:

- Zero-thresholding quantization. Each feature space is partitioned into two quantization intervals based on a global threshold (zero) and a single-bit encoding scheme with two binary labels ("0" and "1") is employed. Zero-thresholding quantization is a special case of static quantization, wherein the threshold is zero. This quantization technique can be formulated by using a simple rule shown in Eq. (7.21).

$$b_i = \begin{cases} 1 & \text{if } T_i^{fl} \geq 0 \\ 0 & \text{if } T_i^{fl} < 0 \end{cases} \tag{7.21}$$

  where T_i^{fl} denotes the ith element of the PGTQ-based template T^{fl} and b_i represents the ith binary value corresponding to the ith element of T^{fl}.
- Reliability-based dynamic quantization. In contrast to the static approach, dynamic quantization approach assigns a variable number of allocated bits to each feature component according to the distinct discriminability of the user-specific feature. In this chapter, an instance of dynamic quantization, namely reliability-based dynamic quantization proposed by Lim *et al.* [20] is adopted. The procedure of this method is briefly described as follows:

  1. Statistical analysis
     Let I be the number of users who participated in the training process, N^t be the number of training samples of each user and T^{fl} denote the real-valued fixed-length representation.

Step 1.1: Compute the mean feature vector μ_i^{fm} of user i and the grand mean vector μ^g.

$$\mu_{id}^{fm} = \frac{1}{N^t}\sum_{j=1}^{N^t} T_{ijd}^{fl}$$
$$\mu_i^{fm} = [\mu_{i1}^{fm} \dots \mu_{iN_{\text{dim}}}^{fm}] \tag{7.22}$$

$$\mu_d^g = \frac{1}{I \cdot N^t}\sum_{j=1}^{N^t}\sum_{i=1}^{I} T_{ijd}^{fl}$$
$$\mu^g = [\mu_1^g \dots \mu_{N_{\text{dim}}}^g] \tag{7.23}$$

where j and d denote the jth sample of user i and the dth dimension of T^{fl}, respectively.

Step 1.2: Compute the within-class variance vector v_i^{wc} of the ith user and the between-class variance vector v^{bc}.

$$v_{id}^{wc} = \frac{1}{N^t}\sum_{j=1}^{N^t}\left(T_{ijd}^{fl} - \mu_{id}^{fm}\right)^2$$
$$v_i^{wc} = \left[v_{i1}^{wc} \dots v_{iN_{\text{dim}}}^{wc}\right] \tag{7.24}$$

$$v_d^{bc} = \frac{1}{I}\sum_{i=1}^{I}\left(\mu_{id}^{fm} - \mu_d^g\right)^2$$
$$v^{bc} = \left[v_1^{bc} \dots v_{N_{\text{dim}}}^{bc}\right] \tag{7.25}$$

Step 1.3: Calculate the signal-to-noise ratio vector ε_i of the ith user.

$$\varepsilon_i = \left[\varepsilon_{i1} \dots \varepsilon_{iN_{\text{dim}}}\right] = \left[\frac{v_1^{bc}}{v_{i1}^{wc}} \dots \frac{v_{N_{\text{dim}}}^{bc}}{v_{iN_{\text{dim}}}^{wc}}\right] \tag{7.26}$$

2. Reliability weight computation
According to equal background probability mass of normal distribution, the feature space is quantized into 2^n intervals. Each interval is labeled with an n-bit gray code. With the fixed-length

representation T^{fl} obtained from Section 7.3.2, the following conversion is performed:

Step 2.1: Each real-valued component of T^{fl} is mapped to an n-bit gray code segment according to the interval labels:

$$\mathrm{GC}(T^{fl}_{ijd}) = [b^1_{ijd} \cdots b^n_{ijd}] \tag{7.27}$$

where b^x_{ijd} denotes the xth bit allocated to the dth feature component for the jth sample of the ith user.

Step 2.2: Compute the reliability weight vector w_d for each binarized feature component by counting the agreeing bits at each bit position and normalizing each summation.

$$\begin{aligned} w_d &= \{w_{d1}, w_{d2}, \ldots, w_{dn}\} \\ &= \left\{ \sum_{i=1}^{I} b_{id1}/I, \sum_{i=1}^{I} b_{id2}/I, \ldots, \sum_{i=1}^{I} b_{idn}/I, \right\} \end{aligned} \tag{7.28}$$

Step 2.3: Repeat steps 2.1 and 2.2 for all $d \in [1, N_{\text{dim}}]$. Then, concatenate all w_d to obtain a user-specific reliability weight vector.

3. Bit allocation based on the weighted reliability and sorted feature.

   Step 3.1: Sort the feature vector in descending order according to the signal-to-noise ratio ε_i obtained in step 1.3. The sorted indices of the feature components are stored as

$$[s_i(1)\, s_i(2), \ldots, s_i(N_{\text{dim}})] \leftarrow \text{sort}_{\text{desc}}(\varepsilon_i) \tag{7.29}$$

   Step 3.2: In order to select the M_{rb} most reliable bits from the extracted N_{dim} bits string, a reliability weight threshold τ_{rw} is used as follows:

$$\tau_{rw} = 1 - N_{\text{step}} \tag{7.30}$$

   where N_{step} denotes an incremental variable ranging from 0 to 0.5 with a sorting step of $N_{\text{step}} = \frac{1}{N^t}$. The bit allocation process is described as follows: Starting from the first sorted feature component, if the reliability of the jth bit satisfies the threshold τ_{rw}, this bit is selected and the evaluation proceeds with the next bit of the same feature component. If any bit fails to satisfy the threshold, the subsequent bits of the same feature component will be skipped for selection evaluation and the evaluation proceeds with the next sorted feature component. If after the evaluation of the last sorted feature component, the selected bits are less than the required length of binarized representation of T^{fl}, the evaluation restarts with a lower step of reliability threshold and is carried out on the bits that are skipped during the previous rounds according to the sorted order of the feature components.

Step 3.3: Once M_{rb} most discriminable and reliable bits have been selected, the bit allocation is terminated. The final binary representation with length

$$M_{rb} = \sum_{i=1}^{N_{\text{dim}}} m_{id}^{b} \tag{7.31}$$

is contributed by the variable number m_{id}^{b} of selected bits from every dth feature component of the user, where $m_{id}^{b} \in \{1, \ldots, n\}$. For every dimension d, the cutpoints of $2^{m_{id}^{b}}$intervals and the m_{id}^{b} values are stored as the helper data.

7.3.4. *Matching two fixed-length representations*

Two types of representation for fixed-length matching, real-valued and binary, are considered.

- [Real-valued]: Let $T_k^{flr,e}$ and $T_k^{flr,q}$ be two fixed-length real-valued representations. The matching score is calculated as

$$S_r^{fl} = \frac{\sum_{k=1}^{N_{\text{dim}}} \left(T_k^{flr,e} * T_k^{flr,q}\right)}{\sum_{k=1}^{N_{\text{dim}}} \left(T_k^{flr,e}\right)^2 + \sum_{k=1}^{N_{\text{dim}}} \left(T_k^{flr,q}\right)^2} \tag{7.32}$$

where $*$ represents an element-wise multiplication operator; $\sum_{k=1}^{N_{\text{dim}}} (T_k^{flr,e})^2$ and $\sum_{k=1}^{N_{\text{dim}}} (T_k^{flr,q})^2$ denote the summation of squared elements of $T_k^{flr,e}$ and $T_k^{flr,q}$, respectively.

- [Binary]: Let $T_k^{flr,e}$ and $T_k^{flr,q}$ be the two fixed-length binary representations. The matching score is computed as

$$S_b^{fl} = 1 - \frac{\sum_{k=1}^{M_{\text{bit}}} \left(T_k^{flr,e} \oplus T_k^{flr,q}\right)}{M_{\text{bit}}} \tag{7.33}$$

where $\oplus$ represents a bit-wise XOR operator; M_{bit} denotes the bit-length of the binary representations $T_k^{flr,e}$ and $T_k^{flr,q}$.

7.4. Point-to-String Conversion Based on Bag-of-Minutiae

The concept of bag-of-minutiae (BoM) originates from the bag-of-words (BoW) model for text classification [27], where a text, may it be a sentence, paragraph or document, is classified by the occurrence of words from the dictionary in the text (example in Fig. 7.2). From the perspective of data representation, BoW modelling transforms a text into a histogram summarizing the frequency of words used in the text. Moving one step closer to the representation of fingerprints, the BoW model has also been implemented in many image classification problems [12, 31], in which it was renamed as bag-of-visual-words (BoVW). Here, visual words refer to local feature descriptors generated based on the keypoints on the image (example in Fig. 7.3). Some commonly used keypoint extraction methods in image classification include Laplacian of Gaussian (LoG), scale-invariant feature transform (SIFT), Hessian Laplace and Hessian Affine.

Since each written word has a unique formation of characters or symbols, the dictionary for BoW is simply composed of all distinct words up to date and two words are said to be correspondences if the formations match with each other perfectly. On the other hand, two corresponding keypoints on two different images might not produce exactly the same descriptor. Therefore, a set of predefined descriptor

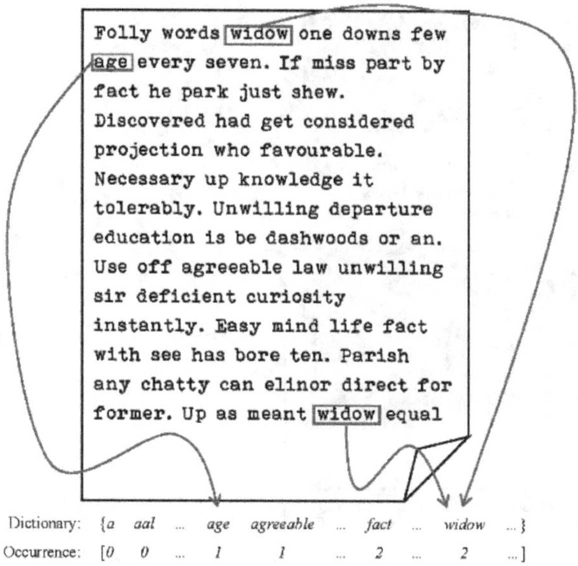

Fig. 7.2. BoW recognizes a text by observing the frequency of words from the dictionary appearing in the text.

prototypes is needed to make up the dictionary so that the descriptors on the query image can be mapped onto the nearest prototypes in the dictionary to generate a histogram similar to that of BoW. Considering the minutiae extracted from fingerprints as a type of keypoint, the BoW equivalent of minutiae-based fingerprint representation is a histogram capturing the appearance of certain predefined minutia prototypes in a fingerprint, thus the name bag-of-minutiae. The essence of introducing BoM is that it is able to transform the minutiae representation (local features) of fingerprints into a globally fixed-length histogram or histogram-like vector.

7.4.1. *Approaches to minutiae-based fingerprint matching using BoM or BoM-Akin methods*

There are several minutiae to fixed-length vector transformation methods proposed in the literature that yield the characteristics of, or are similar to, the BoM modelling.

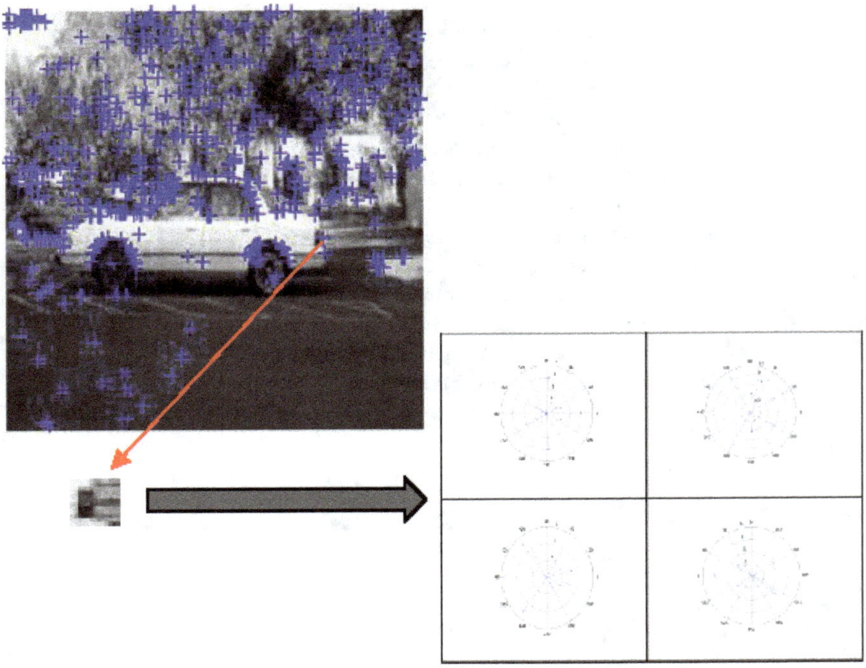

Fig. 7.3. The blue crosses represent the SIFT keypoints extracted from the sample image. For each keypoint, a visual word can be constructed based on the local neighbourhood region. In this case, a SIFT descriptor is a set of orientation histograms created based on the orientations of the region.

Ahmad *et al.* [2] projected the minutiae onto a randomly created straight line and divided the line into multiple regions to obtain a fixed-length vector. Sutcu *et al.* [34] used a local point aggregation approach which constructs random cuboids on the three-dimensional space (x-coordinate, y-coordinate and orientation) and produces a histogram based on the number of minutiae enclosed in each cuboid. This work was extended by Nagar *et al.* [23], in which more discriminative features were used for binning, such as the distance from minutiae to the nearest boundary, the average and standard deviation of minutiae coordinates. Besides, highly correlated bits are eliminated to improve the performance. Moreover, Gudkov and Ushmaev [8] employed random minutiae cluster selection and performed matching on the fixed number of selected minutiae only. However, this method suffers great loss in accuracy due to the possibility of indistinctive minutiae being chosen. These histogram-based methods operate on the spatial domain of minutiae, which directly utilizes the ISO template of minutiae in S2V transformation, and thus require fingerprint pre-alignment. Moreover, Bringer and Despiegel [4] proposed a method to generate the binary fingerprint feature vector from minutiae vicinities, in which each minutia is represented by a vicinity vector. These vicinities are then matched against a set of N representative vicinities. An N-bit binary string is acquired by thresholding the maximum score for each representative vicinity. Another approach to convert the minutiae set into a fixed-length binary template was introduced by Farooq *et al.* [6]. Initially, the attributes of each minutiae triplet are converted into an N-bit string. The final bit-string is generated by identifying the triplet patterns which only appear once in the fingerprint out of the histogram of length $2N$. Similarly, Sandhya *et al.* [28] performed the Delaunay triangulation based on the minutiae, and used the edges and angles of the triangles as the invariant features. The feature dimension was then quantized into cells with predefined size to form a bit-string that registers the cells containing minutiae. A more sophisticated technique, namely the k-means algorithm was used by Vij and Namboodiri [36] to determine the representative features in order to attain better performance. The experimental results of the works discussed above are summarized in Table 7.1.

7.4.2. *Design of the BoM model*

Bianconi and Fernandez [3] discussed the five dichotomous design choices of BoW modelling for image classification but not all are applicable in the context of minutiae-based fingerprint recognition. The three design approaches of the BoM model, each with two alternative operational modes as illustrated in Fig. 7.4, are explained here as a modification to Bianconi and Fernandez's framework. Also, the methods reviewed above are categorized according to the modified framework as shown in Table 7.2.

Table 7.1. Experimental results of existing BoM or BoM-akin methods represented in EER.

References	EER
[2]	≈ 6% (FVC2002 DB2)
[34]	2.7% (Mitsubishi)
[23]	2% (FVC2002 DB2)
[8]	15% (NIST SD14)
[4]	1.7% (FVC2002 DB2)
[6]	1.59% (FVC2002 and FVC2004)
[28]	3.96% (FVC2002 DB1); 2.98% (FVC2002 DB2); 6.89% (FVC2002 DB3); 12.17% (FVC2004 DB1); 13.29% (FVC2004 DB2); 17.73% (FVC2004 DB3)
[36]	≈ 1% (FVC2002 DB1, DB2 and DB3); ≈ 8% (FVC2004 DB1 and DB2)

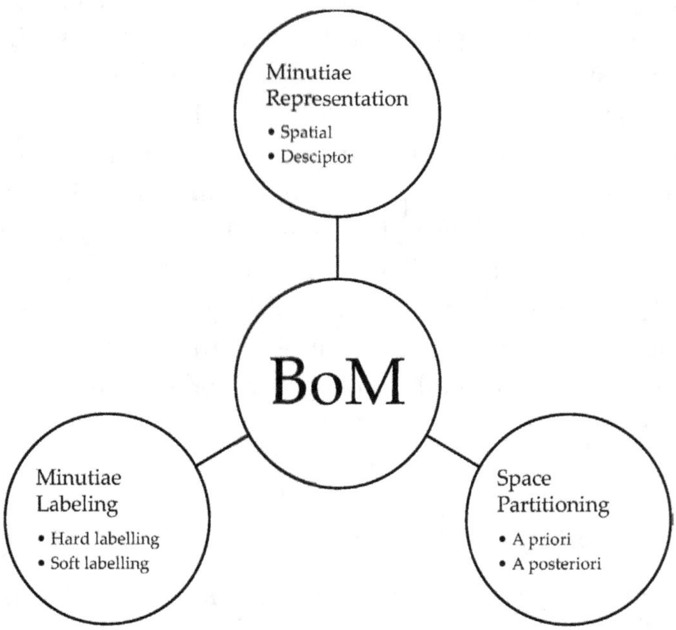

Fig. 7.4. Three design choices of the BoM model.

7.4.2.1. *Minutiae representation*

The choice of minutiae representation refers to the feature space which BoM operates on. It can be either the three-dimensional spatial representation or the descriptor-based representation. The former

Table 7.2. Summary of existing fixed-length vector transformation methods categorized according to the BoM design framework.

Reference	Minutiae representation	Space partitioning	Minutiae labelling
[2]	S	APr	HL
[34]	S	APr	HL
[23]	S	APr	HL
[8]	D	APr	HL
[4]	D	APo	HL
[6]	D	APo	HL
[28]	D	APr	HL
[36]	D	APo	HL

Note: S: spatial representation; D: descriptor representation; APr: *a priori partitioning*; APo: *a posteriori partitioning*; HL: hard labelling; SL: soft labelling.

directly utilizes the ISO template of minutiae (except for the minutia type) in space partitioning, whereas the latter transforms minutiae into the descriptor space prior partitioning.

The spatial representation of a minutia indicates the location of the minutia on the fingerprint image and its ridge orientation. The minutiae aggregation approach [23, 34] and the projection-based approach [2] are examples of creating clusters based on this representation. Since translation and rotation often occurs in fingerprint images and are unpredictable, the drawback of using such representation is that the partitions should move accordingly, and thus are impossible to define. One solution to this issue is to perform fingerprint pre-alignment so that the fingerprints are properly aligned with the space where the partitions are constructed on as shown in Fig. 7.5. Although the vast advancement in the current computer technology has set aside the concern over a high computational time of fingerprint pre-alignment, the errors in such process can affect the accuracy of the transformed fingerprint template.

On the other hand, the descriptor-based representation transforms the minutiae into a translation- and rotation-invariant local space and consequently eliminates the need for fingerprint prealignment. Descriptors employed for set-to-vector transformation in the literature are the fixed-radius vicinity [4] and minutiae triplets [6, 36], among which the fixed-radius descriptor is more robust

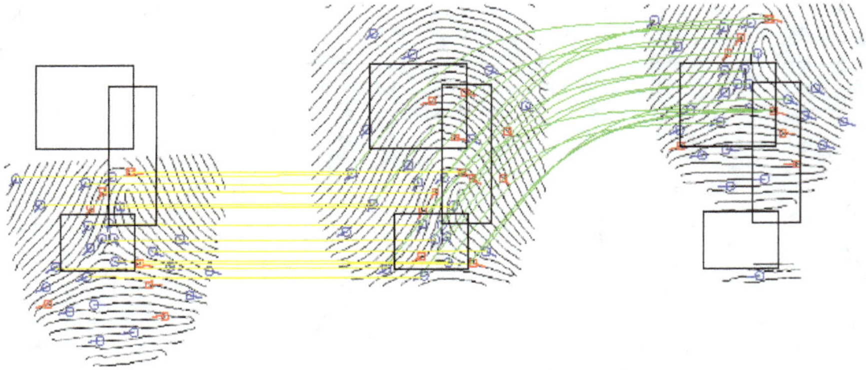

Fig. 7.5. An example showing the effect of (with and without) fingerprint pre-alignment on minutiae aggregation approach similar to that proposed by Sutcu *et al.* [34] and Nagar *et al.* [23]. The fingerprint in the middle is registered and random rectangles are constructed as BoM partitions; the fingerprint on the right is a query image; and the fingerprint on the left is the query image after pre-alignment. Note that if the partitions remain at the same spatial locations, the regions enclosed by the partitions are greatly translated without fingerprint pre-alignment and hence do not produce similar minutiae counts as the registered ones. On the opposite side, pre-aligned image is able to generate similar results as the registered image except for errors caused by missing and/or spurious minutiae.

against local non-linear distortions. From Table 7.1, it is understood that the descriptor-based representation outperforms spatial representation.

7.4.2.2. *Space partitioning*

Two alternative options of partitioning the feature space include *a priori* and *a posteriori*. *A priori* partitioning is independent of data and the resulting partitions are usually user-defined. For example, the random cuboid clustering technique [23, 34] creates randomly chosen partitions of different sizes on the feature space regardless of the minutiae distribution. Equal-width quantization of the feature space [28] is another *a priori* partitioning technique. As the minutiae are not uniformly distributed, redundant and overpopulated partitions are unavoidable using these methods.

On the contrary, *a posteriori* partitioning requires the knowledge of data distribution and the partitions are learnt from the preliminary data. The clusters selection technique used by Bringer and

Despiegel [4] and Farooq *et al.* [6] demonstrates *a posteriori* partitioning, in which cluster centroids are selected from the training data through certain criteria. A more advanced technique was used by Vij and Namboodiri [36] where the partitions are learnt through the K-means algorithm. Other algorithms such as various supervised and unsupervised learning algorithms are also possible approaches towards *a posteriori* partitioning. *A posteriori* partitioning has been proven to deliver better recognition performance than its counterpart according to Table 7.1.

7.4.2.3. *Minutiae labelling*

Minutiae labelling refers to the process of assigning minutiae to the most appropriate partition(s). The options are hard labelling and soft labelling. The former assigns each minutia to the nearest partition only, while the latter allows a minutia to be associated with multiple partitions. All existing methods [4, 6, 23, 28, 34, 36] adopt the concept of hard labelling.

Besides K-means clustering used by Vij and Namboodiri [36], other possible hard labelling techniques include K-means++, K-medians clustering and a self-organizing map. The output of these labelling techniques is binary | a minutia either belongs to a partition or not. Although computationally simple, hard labelling is sensitive to noisy data and outliers. Besides, it is not able to handle two or more highly overlapping classes. As opposed to that, soft labelling introduces fuzziness to the partition borders so that close-to-border minutiae may be assigned to more than one partition. By introducing such tolerance in minutiae assignment, the quantization error may be reduced. Soft labelling usually produces a real-numbered weight indicating the strength or probability of a minutia belonging to a partition as the output. As the minutiae obtained for a fingerprint are different each time, the fingerprint is scanned due to translation, rotation and other skin conditions, such a weight value helps reduce the error caused by mislabelling a minutia when distortions occur. A few instances of soft labelling include fuzzy C-means clustering, sparse approximation and Gaussian mixture modelling. There has not yet been any example in the literature that uses soft labelling for transforming minutiae into a bit-string.

7.4.3. *Towards realizing soft minutiae labelling with a posteriori partitioning using descriptor-based minutiae representation*

From Section 7.3 and Table 7.1, we can see that descriptor-based minutiae representation is superior to spatial representation, and *a posteriori* partitioning has performance advantage over *a priori* partitioning. However, soft labelling has not been adopted in the existing methods. Therefore, we would like to introduce a fixed-length vector transformation that combines these three elements.

In order to qualify for *a posteriori* method, the transformation model should have prior knowledge of the data distribution. In this case, the data refers to the minutiae vectors generated using any minutia descriptor. Learning from the data distribution, the model should define the partitions on the feature space so that similar minutiae are grouped together while distinct minutiae are well separated. This can be done through the training stage. Given a set of training data $\boldsymbol{M}_t \epsilon \mathbb{R}^{D_m \times N_t}$ containing N_t minutiae vectors of D_m dimensions, the *a posteriori* learning task can be translated into the equation below:

$$\min_{\boldsymbol{D},\boldsymbol{V}} \|\boldsymbol{M}_t - \boldsymbol{D}\boldsymbol{V}\|_F^2 \tag{7.34}$$

where $\boldsymbol{D} \in \mathbb{R}^{D_m \times K}$ is known as the dictionary and $\boldsymbol{V} \in \mathbb{R}^{K \times N_t}$ is the label matrix. The dictionary contains K basis vectors and each basis vector corresponds to a partition.

For hard labelling, each column of the label matrix consists of only one non-zero element to indicate that one minutia vector belongs to one partition. On the contrary, the label matrix produced by soft labelling has more than one non-zero elements in each column. Hence, (7.34) can be rewritten into

$$\min_{\boldsymbol{D},\boldsymbol{V}} \|\boldsymbol{M}_t - \boldsymbol{D}\boldsymbol{V}\|_F^2 \quad \text{s.t. } \forall i, \|\boldsymbol{V}_i\|_0 \leq \lambda \tag{7.35}$$

for soft labelling, where $\|\boldsymbol{V}_i\|_0$ represents the ℓ^0-norm of the ith column of $\boldsymbol{V}$ and $\lambda \in N$ is the maximum number of partitions one minutia vector can be assigned to.

Here, we describe the expectation-maximization (EM) approach of solving the optimization problem in (7.35). The training stage involves an iterative process of minutiae labelling and space partitioning. To be specific, K-SVD [1] is one of the EM algorithms that allows soft labelling and will be discussed here as a proposition to solve (7.35).

Algorithm 7.1 demonstrates the walk-through of obtaining the dictionary via K-SVD. Within one dictionary learning iteration, the K-SVD algorithm consists of K subiterations with each corresponding to one basis vector in the dictionary. At the jth subiteration, the approximated minutiae vectors assigned to the jth basis vector, $\boldsymbol{D}_j$, are removed from the set, i.e. $\boldsymbol{D}\boldsymbol{V} - \boldsymbol{D}_j\boldsymbol{V}_{\sim\Lambda}$, where $\boldsymbol{V}_{\sim\Lambda}$ is a subset of $\boldsymbol{V}$ associated with $\boldsymbol{D}_j$. The labelling error matrix is computed as in line 8. SVD then comes into play to update the relevant basis vector as well as label matrix (lines 9–11). Note that the orthogonal matching pursuit (OMP) method is chosen as the labels approximation algorithm here. The iteration repeats itself until the specified maximum number of iterations is met or when the approximation error is less than the threshold.

Once the dictionary is learnt, given a set of minutiae vectors extracted from an input fingerprint, $\boldsymbol{M}$, we are able to obtain $\boldsymbol{V}$ by performing OMP. The fixed-length vector transformation is finally completed by applying one of the following pooling functions on $\boldsymbol{V}$:

(i) Sum-pooling:

$$H = \left[\sum_{i=1}^{N_m} V_{1,i}, \sum_{i=1}^{N_m} V_{2,i}, \dots, \sum_{i=1}^{N_m} V_{K,i}\right] \tag{7.36}$$

(ii) Mean-pooling:

$$H = \left[\frac{1}{N_m}\sum_{i=1}^{N_m} V_{1,i}, \frac{1}{N_m}\sum_{i=1}^{N_m} V_{2,i}, \dots, \frac{1}{N_m}\sum_{i=1}^{N_m} V_{K,i}\right] \tag{7.37}$$

(iii) Max-pooling:

$$H = \left[\max_{i \in [1,N_m]} V_{1,i}, \max_{i \in [1,N_m]} V_{2,i}, \ldots, \max_{i \in [1,N_m]} V_{K,i} \right] \tag{7.38}$$

7.5. Conclusion

A globally fixed-length feature representation derived from varied size biometric features (i.e. *point-to-string conversion*) has been studied over the past decade. Various approaches have been proposed to attempt achieving the objective, e.g. *globally fixed-length*. However, the proposed approaches remain unsatisfactory, such as poor accuracy performance, registration requirement, etc. Nevertheless, the demands of biometric cryptosystems and conventional cryptographic protocols drive further investigation in this topic.

Observations from the past study suggest that a decent *point-to-string conversion* method requires the following characteristics: (1) Comparable accuracy performance over the original feature; (2) Require minimum feature (e.g. sole fingerprint minutia) while remaining registration-free; (3) Security and privacy assurance that can be associated with the conversions. For instance, in a conversion process, the original feature is required as training sample, which may possibly be leaked. Though, the point-to-string conversion method has been studied, we believe it still has a wide room for improvement in the future. We hope this work can provoke thoughts and discussions in this topic.

Bibliography

[1] Aharon, M., Elad, M. and Bruckstein, A. K-SVD: An algorithm for designing overcomplete dictionaries for sparse representation. *IEEE Trans. Signal Process.* **54**(11), 4311–4322, 2006.

[2] Ahmad, T., Hu, J. and Wang, S. String-based cancelable fingerprint templates. In *2011 6th IEEE Conference on Industrial Electronics and Applications (ICIEA)*, pp. 1028–1033, 2011, Beijing.

[3] Bianconi, F. and Fernandez, A. A unifying framework for LBP and related methods. In *Local Binary Patterns: New Variants and Applications*, Springer, pp. 17–46, 2014.

[4] Bringer, J. and Despiegel, V. Binary feature vector fingerprint representation from minutiae vicinities. In *2010 4th IEEE International Conference on Biometrics: Theory Applications and Systems (BTAS)*, pp. 1–6, 2010.

[5] Charikar, M. S. Similarity estimation techniques from rounding algorithms. In *Proceedings of the 34th Annual ACM Symposium on Theory of Computing*, pp. 380–388, 2002.

[6] Farooq, F., Bolle, R. M., Jea, T. Y. and Ratha, N. Anonymous and revocable fingerprint recognition. In *Proceedings of the IEEE Conference on Computer Vision and Pattern Recognition*, pp. 1–7, 2007.

[7] Gionis, A., Indyk, P. and Motwani, R. Similarity search in high dimensions via hashing. In *Proceedings of 25th International Conference on Very Large Data Bases*, Vol. 99, No. 6, pp. 518–529, 1999.

[8] Gudkov, V. Y. and Ushmaev, O. A topologic approach to user-dependent key extraction from fingerprints. In *20th International Conference on Pattern Recognition (ICPR)*, pp. 1281–1284, 2010, Istanbul.

[9] Jain, A. K., Nandakumar, K. and Ross, A. 50 years of biometric research: Accomplishments, challenges, and opportunities. *Pattern Recognit. Lett.* **79**, 80–105, 2016.

[10] Jayasumana, S., Hartley, R., Salzmann, M., Li, H. and Harandi, M. Kernel methods on the riemannian manifold of symmetric positive definite matrices. In *Proceedings of the IEEE Conference on Computer Vision and Pattern Recognition*, pp. 73–80, 2013.

[11] Jin, Z., Lim, M. H., Teoh, A. B. J., Goi, B. M. and Tay, Y. H. Generating Fixed-Length Representation From Minutiae Using Kernel Methods for Fingerprint Authentication. *IEEE Trans. Syst. Man Cybern.: Syst.* **46**(10), 1415–1428, 2016.

[12] Jiang, Y.-G., Ngo, C.-W. and Yang, J. Towards optimal bag-of-features for object categorization and semantic video retrieval. In *Proceedings of the 6th ACM International Conference on Image and Video Retrieval (ACM)*, pp. 494–501, 2007.

[13] Jin, Z., Teoh, A. B. J., Ong, T. S. and Tee, C. A revocable fingerprint template for security and privacy preserving. *KSII Trans. Internet Info. Syst.* **4**(6), 1327–1342, 2010.

[14] Jin, Z., Teoh, A. B. J., Ong, T. S. and Tee, C. Fingerprint template protection with minutiae-based bit-string for security and privacy preserving. *Expert Syst. Appl.* **39**(6), 6157–6167, 2012.

[15] Johnson, W. B. and Lindenstrauss, J. Extensions of Lipschitz mappings into a Hilbert space. *Contemp. Math.* **26**(189–206), 1, 1984.

[16] Juels, A. and Wattenberg, M. A fuzzy commitment scheme. In *Proceedings of the 6th ACM Conference on Computer and Communications Security*, pp. 28–36, 1999.

[17] Krivokuća, V. *Fingerprint Template Protection Using Compact Minutiae Patterns.* Ph.D. dissertation, The University of Auckland, 2015.

[18] Kulis, B. and Grauman, K. Kernelized locality-sensitive hashing. *IEEE Trans. Pattern Anal. Mach. Intell.* **34**(6), 1092–1104, 2012.

[19] Lee, C. and Kim, J. Cancelable fingerprint templates using minutiae-based bit-strings. *J. Netw. Comput. Appl.* **33**(3), 236–246, 2010.

[20] Lim, M. H., Teoh, A. B. J. and Toh, K. A. An efficient dynamic reliability-dependent bit allocation for biometric discretization. *Pattern Recognit.* **45**(5), 1960–1971, 2012.

[21] Liu, E., Zhao, H., Liang, J., Pang, L., Chen, H. and Tian, J. Random local region descriptor (RLRD): A new method for fixed-length feature representation of fingerprint image and its application to template protection. *Future Gener. Comput. Syst.* **28**(1), 236–243, 2012.

[22] Maltoni, D., Maio, D., Jain, A. and Prabhakar, S. *Handbook of Fingerprint Recognition*. Springer Science & Business Media, 2009.

[23] Nagar, A., Rane, S. and Vetro, A. Privacy and security of features extracted from minutiae aggregates. In *2010 IEEE International Conference on Acoustics Speech and Signal Processing* (*ICASSP*), pp. 1826–1829, 2010.

[24] Nandakumar, K. A fingerprint cryptosystem based on minutiae phase spectrum. In *2010 IEEE International Workshop on Information Forensics and Security*, pp. 1–6, 2010.

[25] Pekalska, E. and Haasdonk, B. Kernel discriminant analysis for positive definite and indefinite kernels. *IEEE Trans. Pattern Anal. Mach. Intell.* **31**(6), 1017–1032, 2009.

[26] Rice, J. *Mathematical Statistics and Data Analysis*. Nelson Education, 2006.

[27] Salton, G. and McGill, M. *Introduction to Modern Information Retrieval*, McGraw-Hill Computer Science Series. McGraw-Hill, 1983. ISBN 9780070544840, https://books.google.com.tw/books?id=7f5TAAAAMAAJ

[28] Sandhya, M., Prasad, M. V. and Chillarige, R. R. Generating cancellable fingerprint templates based on delaunay triangle feature set construction. *IET Biometrics* **5**(2), 131–139, 2016.

[29] Schölkopf, B., Smola, A. and Müller, K. R. Kernel principal component analysis. In *International Conference on Artificial Neural Networks*, pp. 583–588, 1997.

[30] Schölkopf, B., Smola, A. and Müller, K. R. Nonlinear component analysis as a kernel eigenvalue problem. *Neural Comput.* **10**(5), 1299–1319, 1998.

[31] Sivic, J. and Zisserman, A. Video Google: A text retrieval approach to object matching in videos. In *2003 Proceedings Ninth IEEE International Conference on Computer Vision*, pp. 1470–1477, 2003.

[32] Sutcu, Y., Li, Q. and Memon, N. Protecting biometric templates with sketch: Theory and practice. *IEEE Trans. Inf. Forensics Sec.* **2**(3), 503–512, 2007.

[33] Sutcu, Y., Sencar, H. T. and Memon, N. A geometric transformation to protect minutiae-based fingerprint templates. In *Defense and Security Symposium*, pp. 1–8, 2007.

[34] Sutcu, Y., Rane, S., Yedidia, J. S., Draper, S. C. and Vetro, A. Feature extraction for a Slepian-Wolf biometric system using LDPC codes. In *IEEE International Symposium on Information Theory*, pp. 2297–2301, 2008.

[35] Tan, S. Y., Jin, Z., Teoh, A. B. J., Goi, B. M. and Heng, S. H. On the realization of fuzzy identity-based identification scheme using fingerprint biometrics. *Sec. Commun. Netw.* **5**(12), 1312–1324, 2012.

[36] Vij, A. and Namboodiri, A. Learning minutiae neighborhoods: A new binary representation for matching fingerprints. In *Proceedings of the IEEE Conference on Computer Vision and Pattern Recognition Workshops*, pp. 64–69, 2014.

[37] Wang, S. and Hu, J. Alignment-free cancelable fingerprint template design: A densely infinite-to-one mapping (DITOM) approach. *Pattern Recognit.* **45**(12), 4129–4137, 2012.

[38] Wang, S. and Hu, J. Design of alignment-free cancelable fingerprint templates via curtailed circular convolution. *Pattern Recognit.* **47**(3), 1321–1329, 2014.

[39] Wang, S. and Hu, J. A blind system identification approach to cancelable fingerprint templates. *Pattern Recognit.* **54**, 14–22, 2016.

[40] Wu, G., Chang, E. Y. and Zhang, Z. An analysis of transformation on non-positive semidefinite similarity matrix for kernel machines. In *Proceedings of the 22nd International Conference on Machine Learning*, Vol. 8, 2005.

[41] Xu, H., Veldhuis, R. N., Kevenaar, T. A., Akkermans, A. H. and Bazen, A. M. Spectral minutiae: A fixed-length representation of a minutiae set. In *IEEE Conference on Computer Vision and Pattern Recognition Workshops*, pp. 1–6, 2008.

Chapter 8

Platform for Detection of Ultra-DNA Signals (10–18 mol/L) with DNA Probe and Nanomaterials

Xun Wang*,‡ and Ben Hoo Mook†,§

*College of Computer and Communication of Engineering
China University of Petroleum, Qingdao, China
†National Tsing Hua University, Hsinchu
Taiwan, 30013, Republic of China
‡wangsyun@upc.edu.cn
§bhmok@hotmail.com

DNA nanotechnology aims to design and build nanodevices or nanostructures for real-life practice. In this chapter, we introduce some state-of-the-art platform for detection of ultra-DNA signals (10–18 mol/L) with DNA probe and nanomaterials. Specifically, DNA sensing technology is recalled from both the construction of new sensing platforms based on nanomaterials with the aim of improving the signal response and improving the reproducibility of the detected results. After that, some platforms for detection of ultra-low DNA signals by gold nanoparticles, magnetic nanoparticles, 2D nanomaterials are discussed.

8.1. Background

Genosensors are biosensors (Fig. 8.1) in which the bio-recognition element consists of a DNA sequence. These devices combine the DNA receptor which imparts selectivity and a transducer which provides

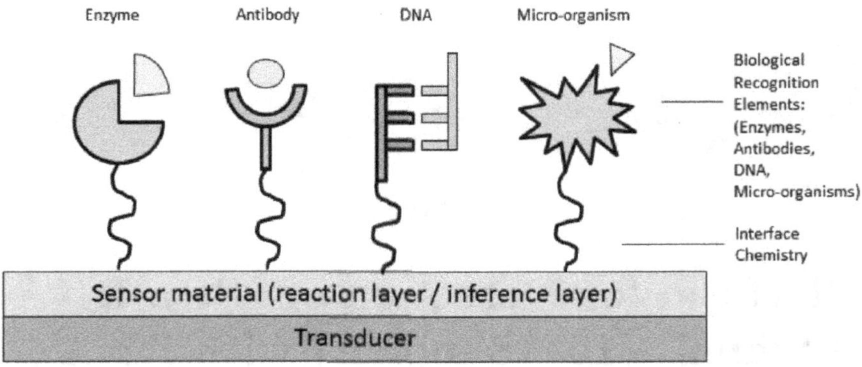

Fig. 8.1. The diagram shows the major elements of a biosensor.

sensitivity and converts the bio-recognition event (the hybridization) into a usable signal. The determination of DNA sequences from humans, animals, bacteria and viruses is the departure point to solve different problems: investigation about food and water contamination caused by microorganisms, detection of genetic disorders, tissue matching, forensic applications, etc. DNA sensing technology has also been widely used in DNA computing, DNA programming and DNA signal detection, etc. In DNA computing and DNA programming, DNA sensors perform a vital function by detecting target signals, which will be regarded as a tool for analysing target expected or unexpected DNA sequences as an information carrier.

Except for the computing purpose or engineering application, DNA or gene signal detection is of great significance in medical examination, intracellular molecular monitoring, and gene disease signal diagnosis, but detection of DNA or gene signals in a low concentration with instant visual results remains a challenge.

8.1.1. *DNA sensing technology*

Among the DNA signal detection sensors, two main groups can be distinguished according to different protocols: based on labelling

DNA target or using a label-free approach. Regarding the first approach, common labels used for hybridization detection can be fluorescent dyes, redox active enzymes, magnetic particles or different kinds of nanoparticles. Label-free electrical detection approach has also recently attracted extensive research efforts since no fluorescent or electrochemical tags are required, which substantially lowers the cost of the detection. In this approach, DNA sensors are based on the detection of unlabelled DNA sequences. This can be performed by measuring the signal due to the direct oxidation of DNA bases or using techniques which are sensitive to changes in the electrical properties of the bio-modified electrode surface, such as Quartz Crystal Microbalance (QMC), Surface Plasmon Resonance (SPR), Electrochemical Impedance Spectroscopy or different kinds of optical spectroscopy.

Meanwhile, many new technologies are being developed to enhance the detection sensitivity of DNA concentration. The Si-nanowire sensors based on the surface-enhanced Raman scattering or field-effect transistor have been demonstrated to detect the DNA at a concentration range of $0.1 \approx 1$ fM. Several reports have recently made advances to approach the detection limit of the subfemtomolar level. An even higher sensitivity in the attomolar range can only be achieved with the aid of signal amplifications.

In the past 10 years, the use of nanomaterials for electrochemical biosensing has seen explosive growth. A wide variety of nanomaterials of different sizes, shapes and compositions are now available. The huge interest in nanomaterials is driven by their exceptional properties. In particular, the ability to tailor the size and structure and hence the properties of nanomaterials offer excellent prospects for designing novel sensing systems and for enhancing the performance of bioanalytical assays.

The unique and attractive properties of nanostructured materials present new opportunities for the design of highly sophisticated electroanalytical DNA biosensing devices. Due to their high surface area, non-toxicity, biocompatibility and charge-sensitive conductance, they act as effective transducers in nanoscale biosensing and bioelectronic devices. This is especially true when the sensed molecules are on the same order of dimension of the used nanocomponents, as this is the case with DNA. These nanostructured material-based electrochemical DNA devices may present a number of key features, including high sensitivity, exquisite selectivity, fast response time and rapid recovery (reversibility), and potential for integration of addressable arrays on a massive scale, which sets them apart from other sensor technologies available today. The sensitivity of the sensor depends on the dimensions and morphological shape of the nanomaterials involved. Therefore, some morphological (nanotube, nanowires, nanofibers, nanorods)-based biosensing transducers could function as effective mediators and facilitate the electron transfer between the active site of the probe DNA and surface of the electrodes.

The use of nanomaterials in genosensing involves two different aspects. Some works focus on the study and construction of new sensing platforms based on nanomaterials with the aim of improving the signal response (i.e. enhancing the sensitivity of the technique or improving the reproducibility of results). Others are based on the use of DNA oligonucleotides labelled with different types of nanoparticles in order to achieve a significant signal amplification.

Two kinds of DNA sensors can be classified according to their application. Electrochemical DNA sensors give high sensitivity, specificity, high selectivity and mutated genes associated with a human disease or low cost for the detection of selected DNA sequences. They also promise to provide a simple, accurate and inexpensive platform for patient diagnosis. Electrochemical DNA sensors also allow sample devices with a minimum size and a very small volume. A field-effect-based DNA sensor is another approach of the application. The use of field-effect-based devices has become a basic structural element in a new generation of biosensors that allow label-free DNA analysis. In particular, ion-sensitive field-effect transistors (FET) are the basis for the development of radically new approaches for the specific detection and characterization of the DNA due to FETs' greater signal-to-noise ratio, fast measurement capabilities and possibility to be included in portable instrumentation. A FET is made by placing metal electrodes on a semiconducting channel. When the electrostatic environment changes around the semiconductor, it changes the conductance of the semiconductor. Label-free biosensors with high miniaturization and integration have attracted intense research interest since they could potentially make advanced low-cost molecular diagnostics routinely available.

8.2. Platform for Detection of Ultra-Low DNA Signals

8.2.1. *Gold nanoparticles*

Gold nanoparticles are expanding many possibilities in labelling and detection in analytical chemistry. Due to the nanoscopic size, gold nanomaterials display novel physical and chemical properties, such as the nanoscale or surface effects. Catalysis is another enhanced feature that can be employed in synthesis or chemically amplified detection. Besides, gold nanoparticles are redox active nanomaterials, that can be electrochemically detected or give way to detection, which make them interesting elements for developing electrochemical biosensors. Gold nanoparticles may improve the sensing properties of the biomolecules and also may enhance the electrón communication rate between redox active species and electrode surfaces. Additionally, gold nanoparticles can be used as both labels and label-free in electrochemical DNA sensing.

Song *et al.* [6] proposed a universal and fast visualized colorimetric method by using gold nanoparticles to detect DNA signals. Specifically, a DNA signal amplification "circuit" based on DNA strand displacement is first developed to amplify the target DNA signals (in low concentrations). After that, thiol-modified hairpin DNA strands and gold nanoparticles are used for visualizing the detection results in a colorimetric manner. If the target DNA signal exits, specific output DNA strands can be produced, with which the thiol-modified hairpin DNA strands and gold nanoparticles cannot combine such that gold nanoparticles aggregate and settle down with colour changing from dark red to grey quickly; if no target DNA signal exits, the gold nanoparticles colloid remains stable and in dark red colour.

Ali *et al.* in [2] reported a methodology for rapid (less than 10 min) and reliable detection of specific sequences and nucleotide mismatch in PCR-amplified and non-amplified DNA using aggregating properties of 40 nm gold nanoparticles as a colorimetric sensor was developed. The assay was label free, required no surface functionalization chemistry, detecting instrumentation and was applicable for species assignment in food analysis. Note that 40 nm gold nanoparticles were prepared by a simple and easily performable method. They demonstrated that the ssDNA was adsorbed on these particles to stabilize them against salt-induced aggregation. The characteristic pinkish-red colour of these colloidal particles was sensitive to the degree of aggregation and was monitored by absorption spectroscopy, transmission electron microscopy or simply by visual observation (Figs. 8.2 and 8.3).

In the research conducted by Song *et al.* it is worth to study the what will happen if situation of the "Threshold" complex constructed in [5] is added to the "circuit." In addition, the concentrations of components are set as μM scale limited by the properties of gold nanoparticles, but more sensitivity is attainable when concentrations are reduced.

8.2.2. *Magnetic nanoparticles*

Magnetic nanoparticles as special biomolecule carriers via a suitable immobilization process offer promise as sensitive sensors. Magnetic nanoparticles have been used in immunoassays and in various reactions involving enzymes, proteins and DNA for magnetically controlled transport and target delivery of anticancer drugs. Their reduced size and ability to be transported in biological systems and reacting medium is an advantage over conventional support systems.

Although nanoparticle-enhanced biosensors have been extensively researched, few studies have systematically characterized the roles

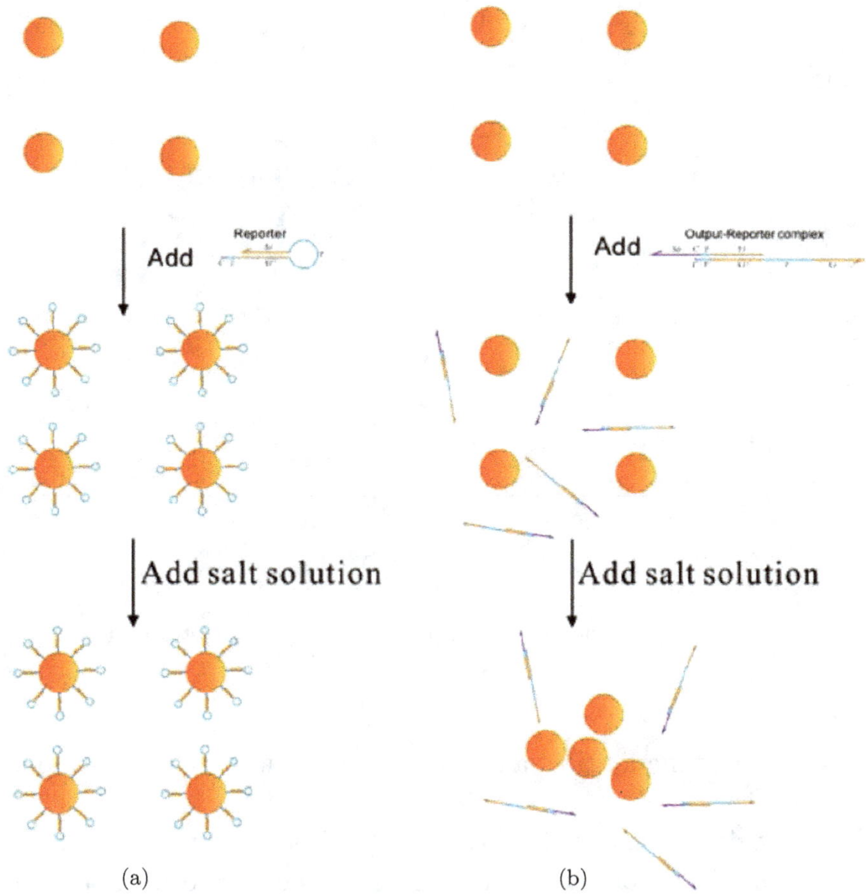

Fig. 8.2. The schematic diagram of the visualized colorimetric DNA signal detection method.

of nanoparticles in enhancing the biosensor functionality. Of all nanoparticles, magnetic nanoparticles are proving to be a versatile tool, an excellent case in point being in DNA bioassays, where magnetic nanoparticles are often used for optimization of the hybridization and separation of target DNA. The use of magnetic nanoparticles, especially Fe_3O_4 ferrites magnetic nanoparticles have been increasing rapidly because they are superparamagnetic, non-toxic and small in size. There have been a number of studies using

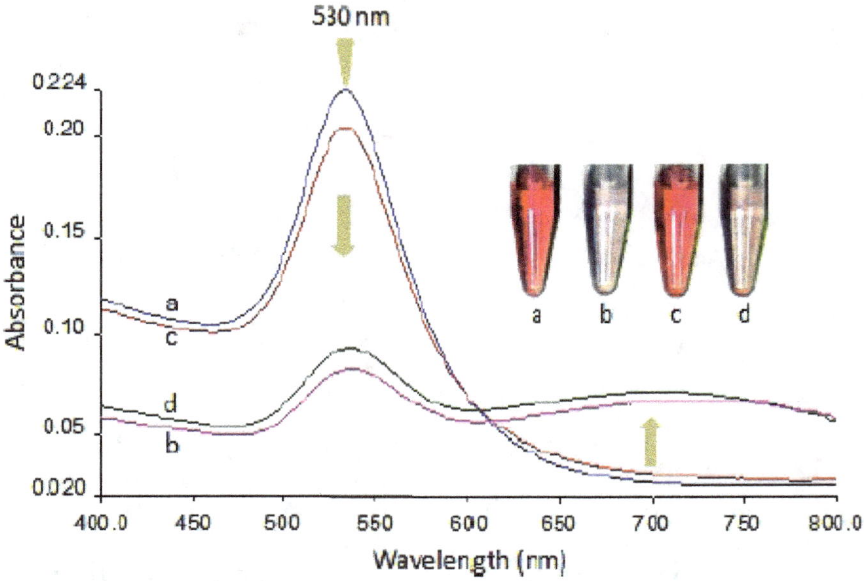

Fig. 8.3. Absorption spectra of aggregated and non-aggregated GNPs.

polymers to functionalize and modify the surface of the magnetic nanoparticles.

Iron oxide magnetic nanoparticles (MNPs) are one of the most important classes of half-metallic materials because of their unique properties. In addition to their high surface to volume ratio, they are highly biocompatible — a property that makes them attractive in biomedicine and research. Furthermore, it is simple to synthesize them and to functionalize their surfaces. This makes them also of interest for use in the fields of information storage and magnetic sensing. The superparamagnetic behaviour of MNPs is revealed when they are a maximum of 20–30 nm in diameter. As a result of this property, in response to a magnetic field they become easily magnetized and demagnetized. If there is an AC field applied, during the demagnetization they release heat due to the loss of energy (specific loss power). This fact has been exploited for the treatment of cancer through hyperthermia.

Tanaka *et al.* [7] propose medical applications using ultra-small magnetic particles and a SQUID magnetic sensor. This system is based on a hybridization process. Two strands in a DNA molecule are held together by hydrogen bonds between base pairs like a ladder. The two strands are referred to as being complementary to each other. One strand (sample DNA) was labelled with Fe_3O_4 ultra-small magnetic particles and the other (probe DNA) was anchored on a glass slide. Then they were hybridized with each other on the slide. After washing the excess sample DNA, the hybridized DNA was evaluated in the presence of excitation AC field by a high-Tc SQUID. The signal was initially proportional to the concentration of the sample DNA and then because saturated. It means that the hybridization occurred successfully between the sample DNA and the probe DNA (Fig. 8.4).

Hushiarian *et al.* in [3] reported a new method of large-scale synthesis of iron oxide (Fe_3O_4) nanoparticles which results in the magnetite particles being in the aqueous phase. Small modifications were applied to design an optical DNA nanosensor based on sandwich hybridization. Characterization of the synthesized particles was carried out using a variety of techniques and CdSe/ZnS core-shell quantum dots were used as the reporter markers in a

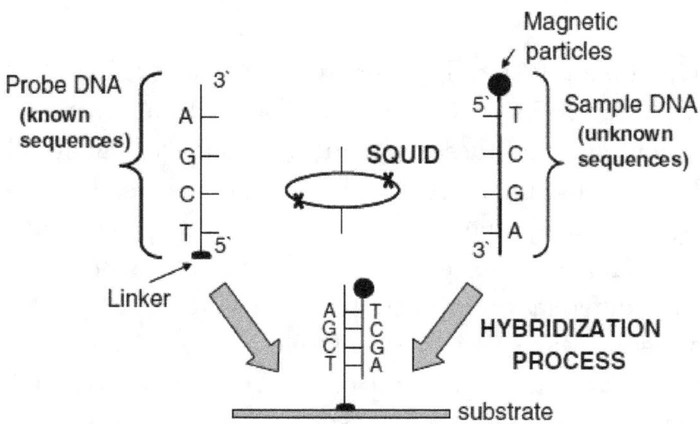

Fig. 8.4. Principle of the detection of DNA sequence.

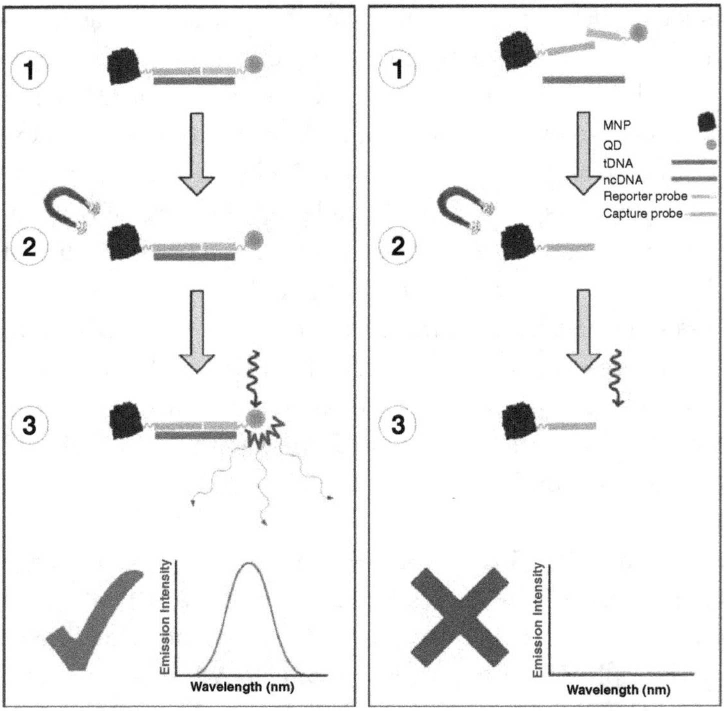

(a) (b)

Fig. 8.5. Schematic diagram of the mechanism of the designed optical DNA nanosensor and the related fluorescence spectra. Three different steps of the process are numbered in the diagram. (a) In presence of complementary tDNA. (b) In presence of non-complementary ncDNA.

spectrofluorophotometer. They showed conclusively that DNA binds to the surface of iron oxide nanoparticles without further surface modifications and that these magnetic nanoparticles can be efficiently utilized as biomolecule carriers in biosensing devices (Fig. 8.5).

Wang *et al.* in [8] reported a GMR sensor-based detection system capable of stable and convenient connection, and real-time measurement. Five different types of MNPs with sizes ranging from 10 to 100 nm were investigated for GMR biosensing. Their experiments were accomplished with the aid of DNA hybridization and detection architecture on the GMR sensor surface and found that different MNPs markedly affected the final detection signal, depending on their characteristics of magnetic moment, size and surface-based binding ability, etc. This work may provide a useful guidance in selecting or preparing MNPs to enhance the sensitivity of GMR biosensors, and eventually lead to a versatile and portable device for molecular diagnostics.

The SQUID system has a large potential to detect small amounts of magnetic nanoparticles because of its high sensitivity for magnetic field. To date, a commercial SQUID system is convenient enough to integrate with other systems. It is interesting to substitute the gold nanoparticles instead of magnetic nanoparticles combined with the SQUID system in the Song research. It may have the opportunity to provide a quick, low cost and low concentration platform in the DNA detection method (Figs. 8.6 and 8.7).

8.2.3. *2D nanomaterials*

Ultra-thin two-dimensional (2D) nanomaterials are attracting dramatically increasing interest after Novoselov, Geim and co-workers

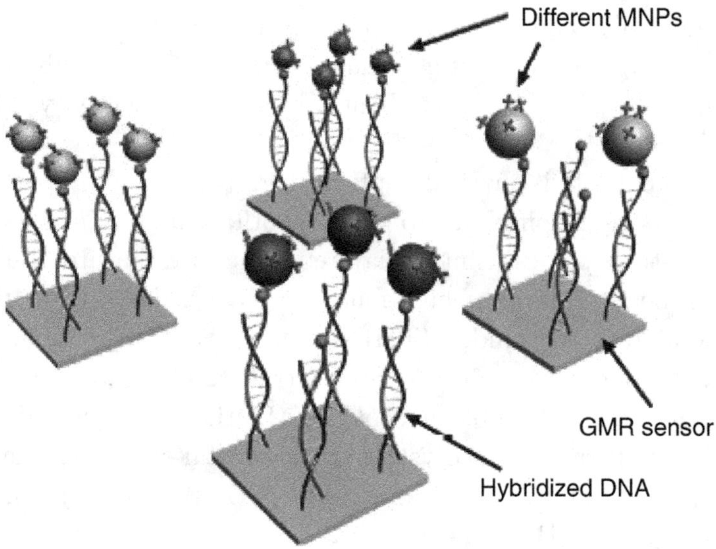

Fig. 8.6. Schematic representation of DNA hybridization on GMR sensors and binding with different MNPs.

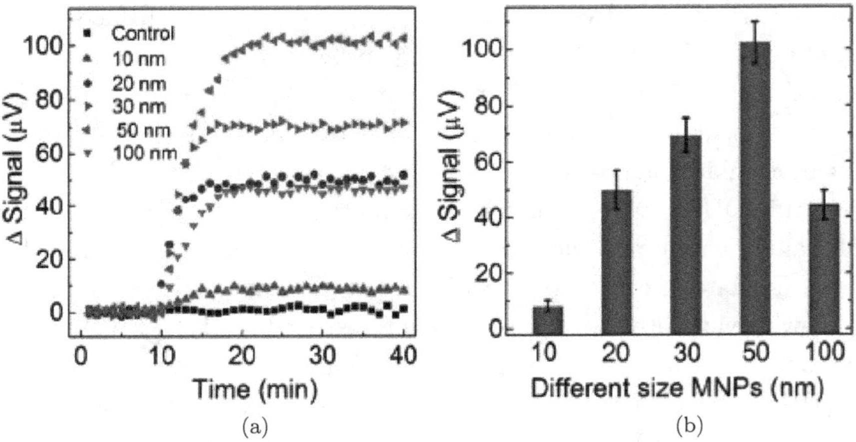

Fig. 8.7. (a) Typical binding curve data in real time and (b) average final signals (with standard deviation) for different MNPs.

exfoliated graphene from graphite using the mechanical cleavage method in 2004. Graphene is a single-atom-thick, crystalline carbon film that exhibits various unprecedented properties, such as ultra-high carrier mobility at room temperature ($\approx$10 000 $cm^2\,V^{-1}\,s^{-1}$), quantum hall effect, large theoretical specific surface area (2630 $m^2\,g^{-1}$), excellent optical transparency ($\approx$97.7%), high Young's modulus ($\approx$1 TPa) and excellent thermal conductivity (3000–5000 $W\,m^{-1}\,K^{-1}$).

The unexpected physical, optical and electronic properties of graphene have inspired us to explore other ultra-thin 2D nanomaterials that possess similar layered structure features but versatile properties, such as hexagonal boron nitride (h-BN), transition metal dichalcogenides (TMDs; e.g. MoS_2, TiS_2, TaS_2, WS_2, $MoSe_2$, WSe_2, etc.), graphitic carbon nitride (g-C_3N_4), layered metal oxides and layered double hydroxides (LDHs). Intriguingly, many new types of ultra-thin 2D crystals, such as metal-organic frameworks (MOFs), covalent organic frameworks (COFs), polymers, metals, black phosphorus (BP), silicene and MXenes, have also been explored in recent years, greatly enriching the family of ultra-thin 2D nanomaterials (Fig. 8.8).

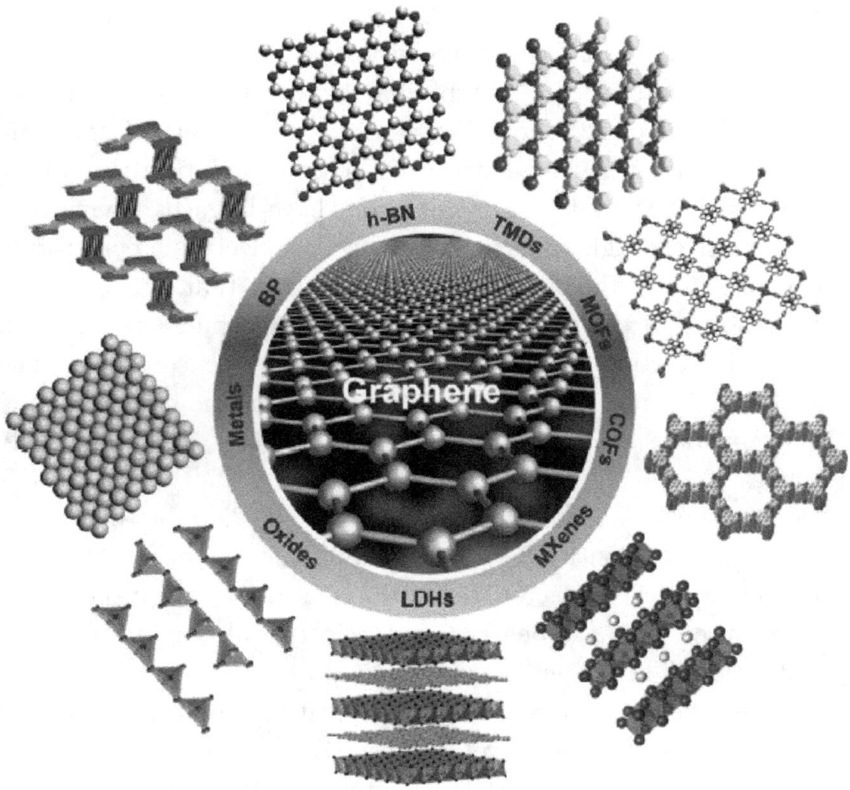

Fig. 8.8. Schematic illustration of different kinds of typical ultra-thin 2D nanomaterials.

The large surface area and unique optical/electrical properties of two-dimensional (2D) monolayer materials make them suitable to interface with biomolecules. For example, grapheme is a highly biocompatible and stable material in aqueous solutions. Graphene and its derivatives graphene oxide have been used to interact with DNA molecules for sensitive detection of DNA hybridization, where the DNA molecules are found to act as potential gating agents that impose *p*-doping to the grapheme layers, leading to the changes in the electrical properties of graphene. In principle, the electrical conduction of an ideal graphene with a Fermi level close to the Dirac point is very sensitive to the external charge perturbation.

Similar to graphene and other 2D materials, 2D MoS_2 offers large surface areas that enhance its biosensing performance [1]. However, due to the existence of a suitable bandgap, the overall sensitivity of devices based on 2D MoS_2 is much larger than that of graphene and graphene oxides which have either no or a small bandgap. Many stoichiometric 2D oxides, in comparison, have large bandgaps that require relatively high applied energy for their electronic band structure modulation. Zhu *et al.* [9] have reported that a single-layer MoS_2 nanosheet exhibits a high fluorescence quenching ability, which assists the detection of biomolecules with fluorescence probes. The ultra-sensitivity of the MoS_2 monolayer to extrinsic charge doping actually provides an excellent avenue for detecting the target DNA molecules.

Phan *et al.* [4] reported a graphene/MoS_2 heterostructure by stacking graphene on a CVD MoS_2 crystalline monolayer, and the photoluminescence characteristics of the constructed graphene/MoS_2 film are used for a label-free and selective detection of DNA hybridization. The graphene serves as a protection layer to prevent the reaction between MoS_2 and the ambient environment as well as a biocompatible interface layer to host DNA molecules on its surface. The photoluminescence (PL) intensity of the MoS_2 layer in

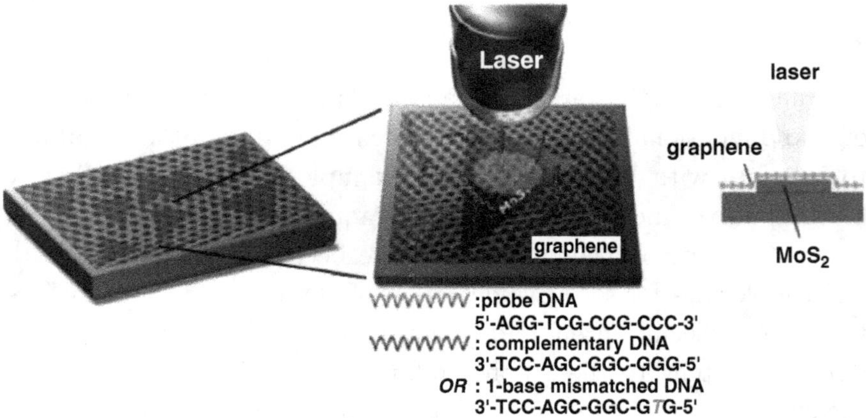

Fig. 8.9. Schematic illustration of the DNA detection method using a microscope and the graphene/MoS_2 heterostructure sensor.

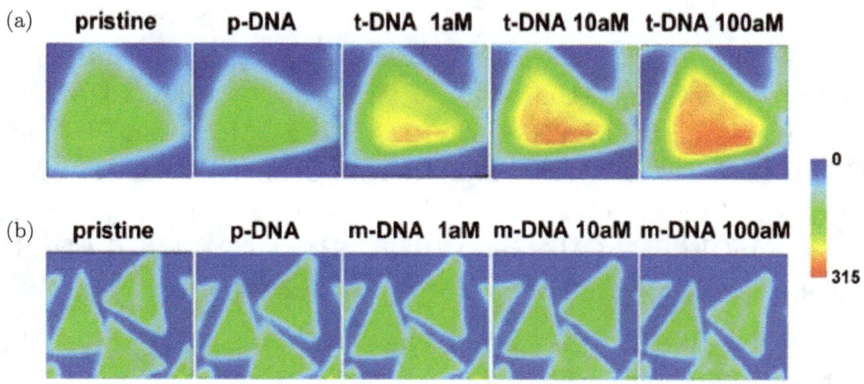

Fig. 8.10. PL peak area mappings of the graphene/MoS_2 heterostructure hybridized with (a) the complementary target DNA and (b) the mismatched DNA.

the graphene/MoS_2 stack increases with the concentration of the added target DNA. The differentiation of complementary and one-base mismatched DNA with the graphene/MoS_2 heterostructure can be performed at a concentration as low as 1 aM (10^{-18} M), indicating high sensitivity for this proof-of-concept heterostructure (Figs. 8.9 and 8.10).

This stacking device shows a great ultra-sensitive platform for a direct attomolar detection. It is anticipated that these kinds of heterostructures may be further expanded to the area of label-free detection of protein, metal-ion contaminants in water, bacteria and intercellular or extracellular activities. These results focus on the sensing of short-chain DNA hybridization (12-mer), and the studies for longer chain DNA or protein molecules are worthy of more investigations. Furthermore, such an extremely high-sensitivity heterostructure is worth further study, especially 2D nanomaterials still leave many alternative materials to study.

Acknowledgments

This work was supported by the National Natural Science Foundation of China (Grant Nos. 61672248, 61572522, 61572523, and 61873280, 61873281), Key Research and Development Program of

Shandong Province (No. 2017GGX10147), Natural Science Foundation of Shandong Province (No. ZR2017MF004), Talent introduction project of China University of Petroleum (No. 18CX02152A), Research Project TIN2016-81079-R (AEI/FEDER, Spain-EU) and Grant 2016-T2/TIC-2024 from Talento-Comunidad de Madrid, Project TIN2016-81079-R (MINECO AEI/FEDER, Spain-EU), and the InGEMICS-CM Project (B2017/BMD-3691, FSE/FEDER, Comunidad de Madrid-EU).

Bibliography

[1] Kalantarzadeh, K., Ou, J. Z. Biosensors based on two-dimensional MoS2. *ACS Sensors* **1**(1), 5–16, 2016.

[2] Ali, M. E., Hashim, U., Mustafa, S., Che Man, Y. B., Yusop, M. H. M., Bari, M. F., Islam, Kh. N. and Hasan, M. F. Nanoparticle sensor for label free detection of swine DNA in mixed biological samples. *Nanotechnology* **22**(19), 195503, 2011.

[3] Hushiarian, R., Yusof, N. A., Abdullah, A. H., Ahmad, S. A. A. and Dutse, S. W. A novel DNA nanosensor based on CdSe/ZnS quantum dots and synthesized Fe_3O_4 magnetic nanoparticles. *Molecules* **19**(4), 4355–4368, 2014.

[4] Loan, P. T. K., Zhang, W., Lin, C.-T., Wei, K.-H., Li, L.-J. and Chen, C.-H. Graphene/MoS_2 heterostructures for ultrasensitive detection of DNA hybridisation. *Adv. Mater.* **26**(36), 4838–4844, 2014.

[5] Shi, X., Wang, Z., Deng, C., Song, T., Pan, L. and Chen, Z. A novel bio-sensor based on DNA strand displacement. *PLoS One* **9**(10), e108856, 2014.

[6] Xin Li, Tao Song, Zhihua Chen, Xiaolong Shi, Congzhou Chen, and Zheng Zhang. A universal fast colorimetric method for DNA signal detection with DNA strand displacement and gold nanoparticles. *J. Nanomater.* **2015**. doi:http://dx.doi.org/10.1155/2015/407184.

[7] Tanaka, S., Aspanut, Z., Kurita, H., Toriyabe, C., Hatuskade, Y. and Katsura, S. Bio-application of high-Tc SQUID magnetic sensor. *J. Magn. Magn. Mater.* **300**(1), e315–e319, 2006.

[8] Wang, W., Wang, Y., Tu, L., Feng, Y., Klein, T. and Wang, J. Magnetoresistive performance and comparison of supermagnetic nanoparticles on giant magnetoresistive sensor-based detection system. *Sci. Rep.* **4**, 5716, 2014.

[9] Zhu, C., Zeng, Z., Li, H., Li, F., Fan, C. and Zhang, H. Single-layer MoS_2-based nanoprobes for homogeneous detection of biomolecules. *J. Am. Chem. Soc.* **135**(16), 5998–6001, 2013.

Chapter 9

Deep Autoencoders in Pattern Recognition: A Survey

Junfen Chen*, Bojun Xie, Hui Zhang and Junhai Zhai

Key Laboratory of Machine Learning and Computational Intelligence
College of Mathematics and Information Science
Hebei University, Baoding, 071002, China
**chenjunfen2010@126.com*

Classification/Recognition performance has a surprising improvement on the prior-art configurations but increases the network depth. However, deeper BP networks are more difficult to train due to gradient diffusion/exploding. From multilayer neural networks to deep neural networks, autoencoders play a significant role to alleviate the gradient diffusion/exploding by means of greedy layer-by-layer unsupervised learning combined with supervised learning. Noisy data and sparsity are introduced into autoencoders to improve the robustness and generalization ability of networks. This study presents an extensive survey on the development of autoencoders used to pretrain deep neural networks, covering three parts: multilayers neural networks and backpropagation learning algorithm; followed by three categories of deep autoencoders; and a review of state-of-the-art literatures about autoencoders in pattern recognition. Overall, an overview about autoencoders is provided in the study.

9.1. Introduction

The history of multilayer neural networks reveals the birth of neural networks in 1958 with perceptrons introduced [38]. The first slow period, referred to as the first AI "winter", has been experienced since the 1970s. During this period, perceptrons could not deal with nonlinear classification/regression, such as XOR problems. Neural networks regained their popularity with the introduction of the backpropagation (BP) algorithm [43]. The widely known view was that representation capability and modelling power increase with an increase in network layers. However, learning in deep architectures using conventional BP algorithm has been proven to be more difficult. Gradient diffusion often occurs in error backpropagation, which leads to failure in training. Stated differently, training a deep network by gradient descent to directly optimize the supervised objective function from random initialization does not perform well. Computers were not sufficiently fast and the dataset was not adequately large, which also constrained the wide applications of neural networks. The second "winter" of neural networks then occurred. This trend changed when studies were conducted on learning processes by the supervised refining process combined with unsupervised pretraining [14, 15]. Neural networks then regained their popularity and have been successfully applied in various fields at remarkable development speed and performance. Deep learning emerged as a widely studied subject in research and modern industries. Specifically, deep architectures may be necessary for tackling challenging recognition tasks to improve generalization performance in machine learning, digital image processing and natural language processing.

With emerging approaches to training deep belief networks (DBNs) and stacked autoencoders, the limitation on learning deep architectures has improved to a certain degree. In this type of approach, a layer-by-layer initialization by unsupervised training is typically used. Initialization is then followed by global fine-tuning of the parameters by using a supervised training criterion. Layerwise, unsupervised learning usually provides an improved initialization to avoid getting entangled in a poor solution. Moreover, the key to unsupervised learning is to yield a good intermediate representation of the inputs [1]. To our knowledge, only several training algorithms can effectively explain what constitutes "good" representations for initializing a deep architecture. These algorithms are restricted Boltzmann machines (RBMs) trained by contrastive divergence (CD) [14, 15], various types of autoencoders [2, 36], semi-supervised embedding [52] and kernel principal component analysis (PCA) [5].

The supervised learning strategy for global fine-tuning combined with unsupervised learning for local pretraining can be used effectively in deep architecture. An empirical study for this strategy is found in the literature [7, 13]. In this study, RBMs and autoencoders have the same deterministic function mapping the inputs to mean hidden representation. Deterministic autoencoders consider a real-valued mean as their hidden representation, whereas stochastic RBMs only sample a binary hidden representation from the mean. A reconstruction error in autoencoders can be regarded as an approximation of the log-likelihood gradient, which is rather similar to the approximation of updates for RBMs by CD. Thus, initializing a deep network by stacked autoencoders yields a classification error almost as low as that by stacked RBMs. Compared with RBMs, autoencoders can avoid complex computation. In addition, autoencoders instead of RBMs produce the best results ever achieved on the most commonly used benchmark for handwritten digit recognition [2]. Thus, this study intends to focus on autoencoders.

The remainder of this study is organized as follows: Perceptrons and the BP algorithm are summarized in Section 9.2, followed by autoencoders. The conclusion and future studies are discussed in Section 9.4.

9.2. Artificial Neural Networks

9.2.1. *Perceptrons*

A perceptron is conceived as a simplified mathematical model to simulate how the neurons in our brains function [38, 39]. The primitive perceptron outputs 1 if the sum of the weighted inputs is larger than the threshold; otherwise, the output is -1. The following perceptrons with a threshold can be explained as perceptrons with a bias input 1, which ensures that more functions can be computed for the same inputs. The latter model was developed on the work of [30]. For example, functions $f(x) = kx$ passing the origin point $(0, 0)$ cannot linearly separate point $(1, 1)$ from point $(2, 2)$; however, the functions with a bias $f(x) = kx + b$ (referring to a translation of the functions) can precisely separate these 2 points. In addition, nonlinear activation functions transfer the weighted inputs to a continuous valued output, which provides considerably wider applications for perceptrons in regression as well as classification. A more general perceptron diagram is plotted in Fig. 9.1.

Prompted by the foundational study [11], Rosenblatt developed a method of learning using the unknown weights $W = [b, w_1, w_2, \ldots, w_n]^T$ for the activation function $f = sign$, namely the perceptron learning algorithm.

Perceptrons achieved the predominant performance on logical reasoning and simple toy shape classification in the early days of AI; regardless, it exhibited a considerably poor performance for nonlinear XOR problems and human handwritten digit classification. The reason is that the input vectors directly connected the output vectors, whereas the input features were manually fixed, rendering perceptrons lacking in internal representation capability for the task domain. Rigorous analysis on the limitations of perceptrons is shown in the study [31] to have resulted in the first AI Winter.

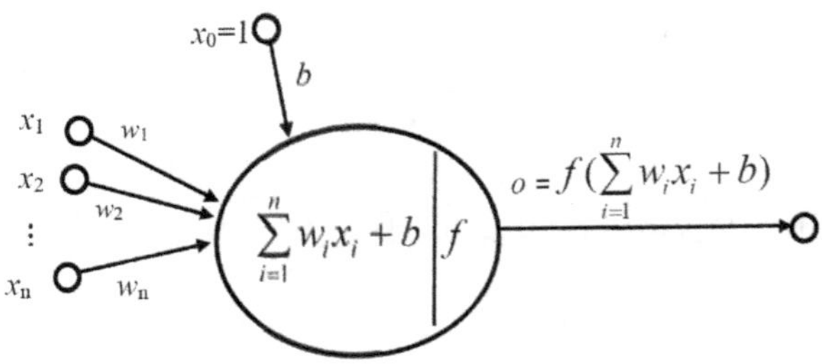

Fig. 9.1. A diagram showing how perceptrons work.

To overcome the limitations of a single perceptron to be applied in nonlinear problems, such as XOR problems, Minsky also proposed to hook multiple perceptrons developing a layer-by-layer fully connected network, namely the artificial neural network (ANN) [31]. Usually, an input layer and an output layer exist, and both of them are determined by the specific task, and at least a hidden layer with a certain number of units is present. A three-layer neural network is plotted in Fig. 9.2. Similar to perceptrons, an extra input with a value of 1 is added to each unit over the hidden and output layers. The weight on this unit is regarded as the bias equivalent to a threshold of the opposite sign. The bias is often treated equally with the other weights. The connection weights are unknown and need to be determined so that the actual outputs of this network are similar to the desired outputs as much as possible for the given samples.

However, the perceptron learning rule cannot work for multiple layers because the method of adjusting the weights of the hidden layer is undetermined. The chain rule can be used to compute the derivatives of neurons and to adjust the weights of hidden layers [51].

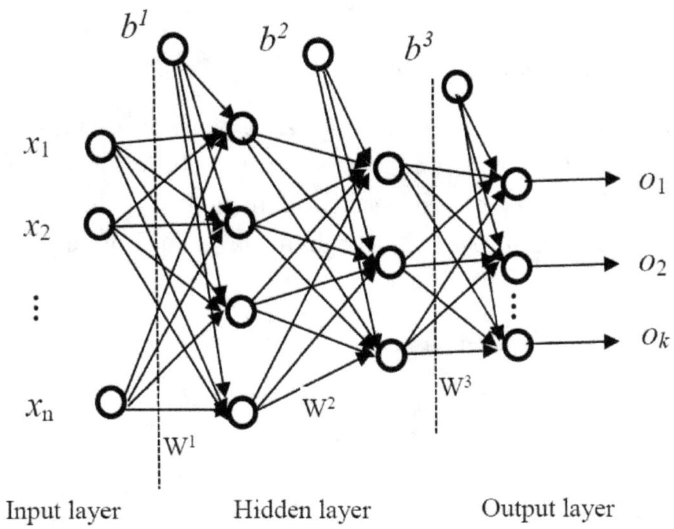

Fig. 9.2. The architecture of a three-layer neural network.

However, the application of BP in neural networks was not published. Until 1986, the BP algorithm, a new learning procedure based on the gradient descent scheme to determine unknown weights, is exactly described in the literature [43]. In the same year, an extended study that specifically addressed problems of perceptrons precisely described the training process of multilayer neural networks [42]. Several more techniques were explored, which enable computers to read the human handwriting. These remarkable advances suggested the return of neural networks.

9.2.2. *Backpropagation learning algorithm*

Conventional backpropagation (BP) algorithm consists of 2 passes: forward propagation information and backward propagation error. The input information is first propagated from the input layer to the output layer. The error is then delivered from the output layer to the input layer before the weights are updated. It repeatedly adjusts the connection weights to minimize the mean squared error between the actual output and the desired output. The main learning procedure of the BP algorithm is described in the following section.

9.2.2.1. *Forward propagation*

A neuron j has an input referred to as x_j and an output y_j, which is usually squashed by a nonlinear activation function, as plotted in Fig. 9.3(b). Here, x_j can be easily identified as a linear combination of the outputs y_i over the left-hand layer and of the connection weights w_{ji}.

Thus, we have

$$x_j = \sum_i y_i w_{ji} \tag{9.1}$$

and

$$y_j = f(x_j) \tag{9.2}$$

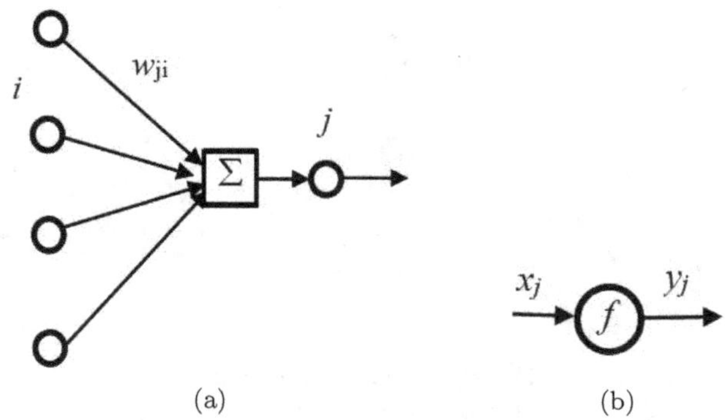

Fig. 9.3. Forward propagation process within a multilayers network. (a) Input weights of neuron j. (b) The input and output of neuron j.

where f is often a sigmoid shape function, for instance, *logsig* function $f(x) = \frac{1}{1+e^{-x}}$, and *tansig* function $f(x) = \frac{e^{x}-e^{-x}}{e^{x}+e^{-x}}$.

When the neuron j lies in the input layer, $y_j = x_j$ is obtained, which may be the reason one counts the number of layers in a network, except for the input layer.

Presently, the information of a given sample has arrived on the output layer; thus, the error E is computed as follows:

$$E = \frac{1}{2}\sum_{j}(y_j - o_j)^2 \tag{9.3}$$

where j is an index over the output units, y is the actual output, and o is its desired output.

9.2.2.2. *Backward propagation*

When the error E is not less than the predefined accuracy threshold, the connection weights are to be modified. The weights between the hidden layer and the output layer can be easily adjusted using the gradient descent scheme because of the computed error over the output layer. Thus, the error for each neuron requires manual computation.

Compared with forward propagation, the input of the neuron j is assigned to e_j, and its output is a_j. When a neuron j is only on the

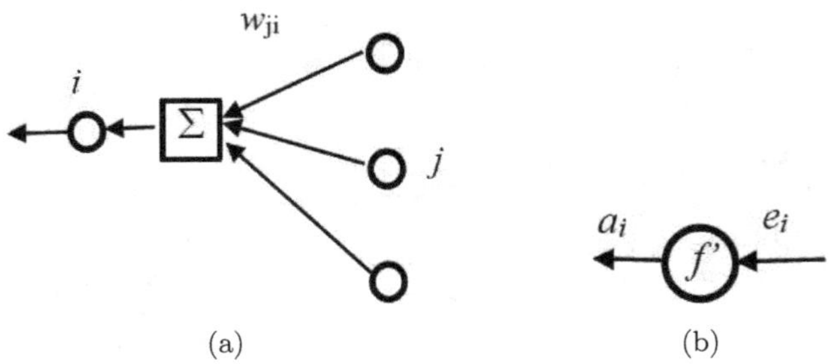

Fig. 9.4. Backward propagation process within a multilayer network. (a) Input connections of neuron i. (b) The input and output of neuron i.

output layer, its input is computed as follows:

$$e_j = \frac{\partial E}{\partial y_j} = y_j - o_j \tag{9.4}$$

On the other hand, when a neuron i is on the hidden layer, its input is merely the sum of the outputs over the right-hand layer (Fig. 9.4(a)), computed as follows:

$$e_i = \sum_j w_{ji} a_j \tag{9.5}$$

When f is appointed to the *logsig* function, the corresponding output a_j therefore is defined as follows (Fig. 9.4(b)):

$$a_j = \frac{\partial E}{\partial y_j}\frac{\partial y_j}{\partial x_j} = y_j(1 - y_j)e_j \tag{9.6}$$

Additionally, from Eq. (9.1), we obtain:

$$\frac{\partial x_j}{\partial w_{ji}} = y_i \tag{9.7}$$

On the basis of the chain rule, the partial derivatives of the error E with respect to the unknown weights are computed as follows:

$$\frac{\partial E}{\partial w_{ji}} = y_i y_j (1 - y_j) e_j \tag{9.8}$$

9.2.2.3. *Weights updating*

Thus, the increment of the weights is proportional to the partial derivative of error E with respect to the weights

$$\Delta w_{ji} = \alpha y_i a_j \tag{9.9}$$

where α is a learning rate within $(0, 1)$ that controls the convergence speed of the gradient descent. Similarly, the increment of the bias is obtained as follows:

$$\Delta b_j = \alpha a_j \tag{9.10}$$

Thus, the weights and biases are updated iteratively according to the following formulas:

$$w_{ji} = w_{ji} + \Delta w_{ji} \tag{9.11}$$

$$b_j = b_j + \Delta b_j \tag{9.12}$$

The BP learning algorithm determines the weights simultaneously, which reveals the ability of hidden units that can achieve useful representation. The ability to create new features significantly distinguishes backpropagation learning from perceptron learning. Kurt proved theoretically that a multilayer BP network is a functional approximator that provides any desired accuracy [17]. Kurt indicated that a 3-layer feedforward network by squashing activation can approximate any Borel measurable function when a sufficient number of hidden units is present. These research findings significantly benefit the following applications in pattern classification or recognition.

9.2.3. *Applications of BP networks in pattern recognition*

Recognition accuracy was markedly improved at the cost of computational consumption caused by highly complex network structures. However, the alternative method of evaluating the generalization ability of BP networks can be greatly enhanced by providing constrains from the task domain. More efficient features of the tasks need to be extracted automatically instead of manually to enhance the recognition performance of BP networks. The BP network was applied to recognize handwritten ZIP code digits by stochastic gradient descent optimization. The inputs in such a method were two-dimensional (2D) normalized images rather than feature vectors [22]. The learning network is directly fed with images with large amounts of low-level information rather than feature vectors. The basic design principle of a network architecture is to minimize the number of free parameters in the network without overly reducing its computational power.

The BP learning algorithm was presented to clearly and concisely address the weight updating procedure for a multilayer neural network [43]. In addition, the initial components of convolutional neural networks, such as receptive fields, weight sharing and feature maps, have been introduced and used to enhance the internal representation of hidden units. A particular feature at any location on the input can be easily detected by weight sharing. Weight sharing refers to several connections controlled by a single weight. Thus, all units in a feature plane share the same set of weights. Weight sharing not only reduces the number of free parameters in the network greatly but also expresses information about geometry and topology of the task.

A neural network-based face detection system was proposed in which small windows on 20×20 input image were examined to detect whether each window contains a face [40, 41]. Receptive fields of the hidden units include 10×10, 5×5 subregions and 20×5 horizontal stripes, which represent different local features. Moreover, a pyramid neural network with a hierarchical architecture was designed for gender classification, in which feature extraction and dimensionality reduction were implemented using local receptive fields in 2D pyramid layers, and features were propagated in one-dimensional (1D) feedforward layers [35]. A face-recognition system using probabilistic decision-based neural networks was proposed, including a face detector, an eye localizer and a face recognizer. This system adopted the hierarchical network structures with nonlinear basis functions and competitive credit assignment to determine the final result [23].

A large deep convolutional neural network was designed by Alex Krizhevsky *et al.* to challenge 1.2 million high-resolution images into 1000 different classes [20]. The neural network, which includes 60 million parameters and 650,000 neurons, consists of 5 convolutional layers, some of which are followed by max-pooling layers, and 3 fully connected layers with a final 1000-way softmax. The effective "dropout" was used to reduce overfitting by randomly omitting half of the feature detectors on each training case [12]. A variant of this model achieved a top-5 test error rate of 0.153, compared with the 0.262 error rate achieved by the second-best in the ImageNet Large Scale Visual Recognition Challenge 2012 competition (ILSVRC2012). The effect of the CNN depth on its accuracy in the large-scale image recognition was investigated by means of exactly exploring receptive fields to improve the performance of networks in another prominent study [46].

9.3. Autoencoders

In the previous section, the mechanism and application of neural networks in supervised learning and conventional backpropagation have been briefly described, with labelled training examples. However, unlabelled data are much easier (and cheaper) than labelled data to collect in real-world applications; thus, unsupervised learning plays a significant role in machine learning. Unsupervised learning can be interpreted as imposing equal constraints among the connection strengths. This technique can be implemented with minimal computational overhead involved.

9.3.1. *A general framework of autoencoders*

An autoencoder commonly consisting of an encoder and a decoder is intended to recover the original input (Fig. 9.5). An autoencoder is an unsupervised learning algorithm, which sets the target values to

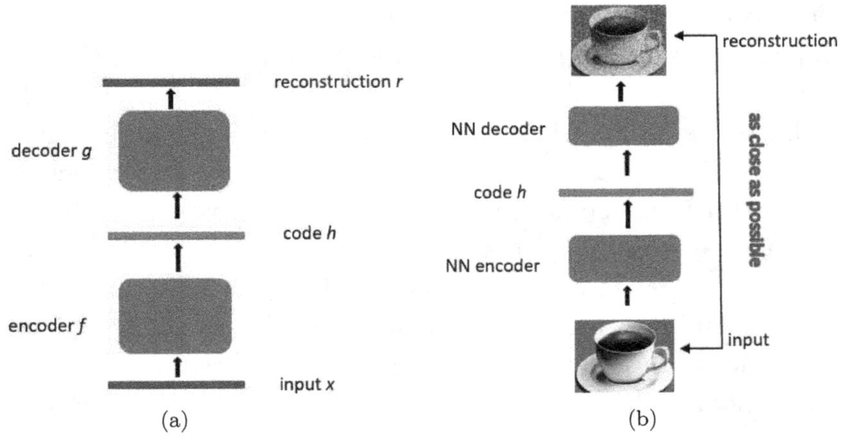

Fig. 9.5. Autoencoders. (a) A general framework of autoencoders. (b) An autoencoder used for image recovery.

be equal to the inputs; that is, $y^{(i)} = x^{(i)}$. An autoencoder applies the BP algorithm to determine unknown parameters, where x is an input and r is an output (a reconstruction of x). Note that f is an encoder for generating $f(x) = h$, and g is a decoder for generating $g(h) = r$. Loss function $L(r, x)$ measures the quality of reconstruction, which is usually defined as the average-squared error or average cross-entropy.

Suppose we have a fixed number of training examples without labels, X, where $x^{(i)} \in R^n$. An autoencoder is designed as a three-layer neural network (Fig. 9.6), which attempts to learn a function $h_{W,b}(x) \approx x$.

In Fig. 9.6, we overload the notations to describe the remainder of this section clearly. $z_i^{(l)}$ and $a_i^{(l)}$ denote the input and output of a unit i, respectively, where $l = 1, 2, 3$. Specifically, $z_i^{(1)} = a_i^{(1)} = x$ for the units in the input layer.

$$z_i^{(l)} = \sum_{i=1}^{s_{l-1}} W_{ij}^{(l-1)} a_i^{(l-1)} + b^{(l-1)}, \quad l = 2, 3 \tag{9.13}$$

$$a_i^{(l)} = f(z_i^{(l)}), \quad l = 2, 3 \tag{9.14}$$

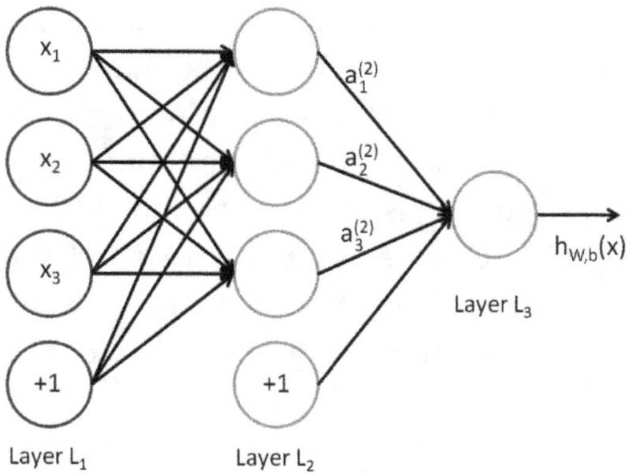

Fig. 9.6. An autoencoder is a three-layer neural network.

where $s_1 = n$ is the number of input units s_2 is the number of hidden units and $s_3 = s_1 = n$. The activation function f often picks up the sigmoid function. Let (W, b) be a collection of parameters; then, the output is $h_{W,b}(x) = a^{(3)}$. Let $W^{(2)} = (W^{(1)})^T$ refer to tied weights. The experimental results demonstrated that $W^{(2)}$ has no need to train; that is, the tied weights setting is valid [7].

The squared error cost function of an autoencoder is defined as follows:

$$J(W, b) = \left[\frac{1}{N}\sum_{k=1}^{N}\frac{1}{2}||h_{W,b}(x^{(k)}) - y^{(k)}||^2\right] + \frac{\lambda}{2}\sum_{l=1}^{2}\sum_{j=1}^{s_l}\sum_{i=1}^{s_{l+1}}(W_{ij}^{(l)})^2. \tag{9.15}$$

The first part is an average sum-of-squared error, and the second part is a regularization term. Regularization reduces the risk of overfitting by decreasing the magnitude of the weights. The parameter λ controls the trade-off between the parts. Standard gradient descent can update the weights W and bias b. Details are provided in the literature [2].

The equation $h_{W,b}(x) \approx x$ can exactly hold by using an identity matrix W and a unit vector b. However, this solution is impractical. To avoid the trivial solution, some constraints need to be added to the autoencoder. These constraints include the number of hidden units, sparsity and tied weights. Different constraints may result in different types of autoencoders. When the number of hidden units is smaller than the number of inputs, an autoencoder is forced to learn a compressed representation of the inputs. Specifically, when a linear activation function is used in the hidden layer, the learned compression is similar to PCA. The limitations of this type of compression include (i) missing information during reconstruction and (ii) the requirement that the testing images be of the same type as the training images.

9.3.2. *Sparse autoencoders*

The discussion in Eq. (9.13) relies on the number of hidden units s_2 being smaller and possibly smaller than the number of input units. However, even in cases opposite that were presented earlier, several interesting structures of the input data can still be discovered by imposing a sparsity constraint on the hidden layer of the autoencoder. A node i is considered active when its output value a_i is close to 1 and is regarded as inactive when a_i is close to 0. The majority of the units are nearly inactive, except for several that respond actively for an input signal. As we know, $a_i^{(2)}(x^{(k)})$ denotes the activation value of the hidden unit i for a given input $x^{(k)}$ in the autoencoder, and

$$\hat{\rho} = \frac{1}{N}\sum_{k=1}^{N}\left[a_i^{(2)}(x^{(k)})\right] \tag{9.16}$$

where $\hat{\rho}$ is the average activation of the hidden unit i over the training examples. We intend to enforce $\hat{\rho} = \rho$, and ρ is a sparsity hyperparameter. ρ is a considerably small positive value (e.g. $\rho = 0.05$ or $\rho = 0.01$).

An extra regularization term is added to the cost function $J(W, b)$, which penalizes $\hat{\rho}$ deviating significantly from ρ. KL divergence $\mathrm{KL}(\rho||\hat{\rho})$ is used to describe this regularization term; thus, the cost function of a sparse autoencoder is defined as follows:

$$J_{\text{sparse}}(W, b) = J(W, b) + \beta KL(\rho||\hat{\rho}) \tag{9.17}$$

where $J(W, b)$ is computed using the formula (9.15), and KL divergence $(\rho||\hat{\rho}) = \sum_{j=1}^{s_2} \left[\rho \log \frac{\rho}{\hat{\rho}} + (1-\rho)\log \frac{1-\rho}{1-\hat{\rho}}\right]$ reaches its minimum of 0 at $\hat{\rho} = \rho$ and increases as $\hat{\rho}$ approaches 0 or 1. Thus, minimizing this penalty term leads $\hat{\rho}$ close to ρ.

The cost function of the sparse autoencoder is rewritten as follows:

$$\begin{aligned} J_{\text{sparse}}(W, b) = &\left[\frac{1}{N}\sum_{k=1}^{N}\frac{1}{2}||h_{W,b}(x^{(k)}) - y^{(k)}||^2\right] \\ &+ \frac{\lambda}{2}\sum_{l=1}^{2}\sum_{j=1}^{s_l}\sum_{i=1}^{s_{l+1}}(W_{ij}^{(l)})^2 \\ &+ \sum_{j=1}^{s_2}\left[\rho \log \frac{\rho}{\hat{\rho}} + (1-\rho)\log \frac{1-\rho}{1-\hat{\rho}}\right] \end{aligned} \tag{9.18}$$

A k-sparse autoencoder was proposed to investigate the effectiveness of sparsity, which only kept k highest activities in each hidden layer [26]. Experimental results showed that this method can achieve better classification performance than those obtained from the denoising autoencoders, networks trained with dropout and RBMs on the MNIST and NORB datasets.

9.3.3. *Stacked autoencoders*

Stacked autoencoders are neural networks consisting of multiple layers of sparse autoencoders in which the outputs of each layer feed the successive layer as inputs (see Fig. 9.7). Stacked autoencoders extend the depth of autoencoders to extract the complex features of the input data further. The network parameters are determined by the common BP algorithm combined with the greedy layerwise mechanism, such as the training process illustrated in Fig. 9.8(a) and 9.8(b). Note that the parameters of each layer are trained individually while parameters for the remainder of the model are frozen.

A stacked autoencoder benefits from any deep network. The first layer of a stacked autoencoder tends to learn first-order features, such as edges in an image. The second layer tends to learn second-order

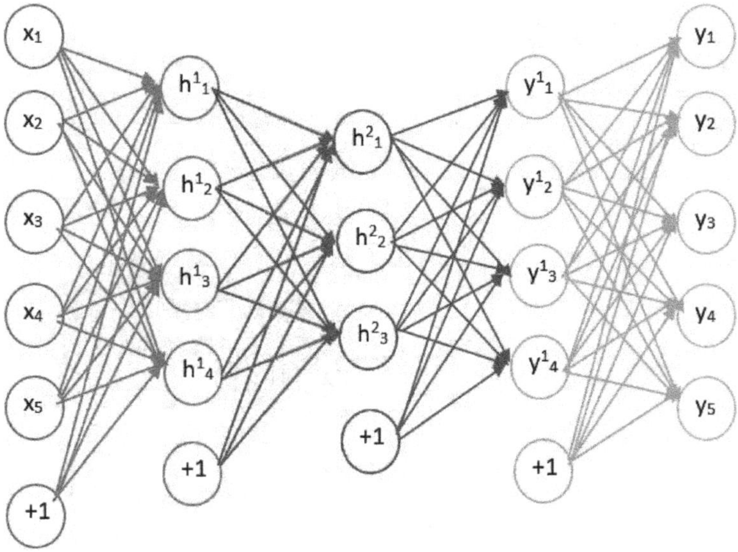

Fig. 9.7. A three-layer stacked autoencoder.

features corresponding to patterns in the appearance of first-order features, such as the contour or corner formed by edges. Higher layers tend to learn even higher order features.

The activation of hidden units in the deepest layer $a^{(n)}$ contains the information that provides a representation of the input in terms of higher-order features. The features from the stacked autoencoder can be used to classify problems by feeding the last hidden layer activation $a^{(n)}$ to a softmax classifier (or any other classifier). The training process of this classifier is illustrated successively in Fig. 9.8(c).

Fine-tuning using the BP algorithm can be used to improve the results by tuning the parameters of all layers simultaneously after the training phase is completed (Fig. 9.8(d)). Figures 9.8(a) and 9.8(b) are clearly identified as unsupervised learning, and Figs. 9.8(c) and 9.8(d) as supervised learning. The effect of fine-tuning heavily relies on the magnitude of the labelled examples among the training data. In the presence of only a small number of labelled data, fine-tuning almost fails to improve the classification performance.

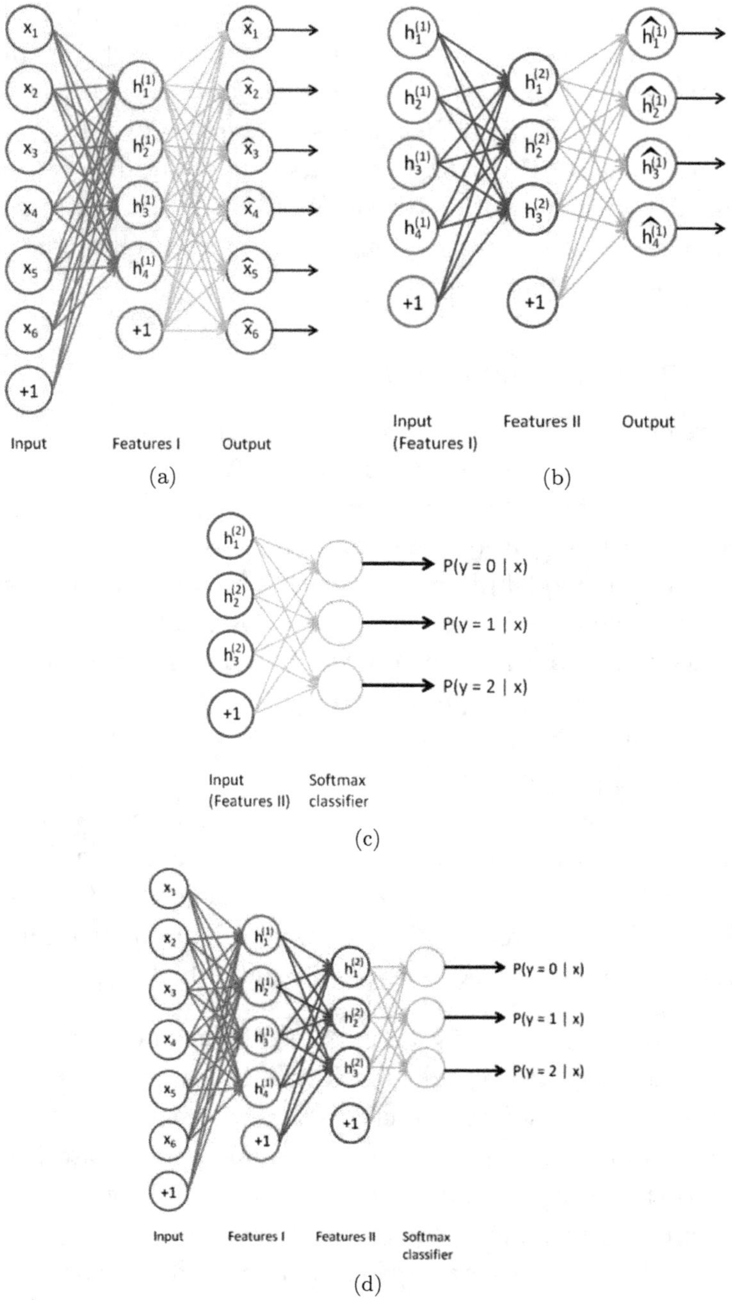

Fig. 9.8. Layerwise training process of the stacked autoencoders in a softmax classifier.

Numerous simple autoencoders are stacked together, implying to a certain degree a deep network, to learn multiple levels of representation and abstraction [21], and layerwise unsupervised learning is merely the core technique that allows a neural network to become widely used again as the third generation of neural-like computing models. This training strategy prompts the frequent use of stacked autoencoders to pretrain deep neural networks to collect the better initialization of the network.

9.3.4. *Denoising autoencoders*

The development of autoencoders is based on the observation that humans have the ability to recognize partially occluded or destroyed images, form a high-level representation associated with different modalities, such as image and sound, and recall the representation even when some modalities are missing. Moreover, a good representation is expected to be stable and robust. Even if the representation is obtained from a partially corrupted input, it can be used to recover the corresponding clean input. The resulting reconstruction criterion called denoising is used in autoencoders to investigate the robustness of the input to partial destruction.

A denoising autoencoder is thus trained to reconstruct a clean "repaired" input from a corrupted version of it [49, 50]. In this study, denoising refers to a training criterion for autoencoders to constitute improved higher level representation. As shown in Fig. 9.9, the training process of a denoising autoencoder is described as follows:

- First, corrupting an initial input x into $\tilde{x}$ by stochastic mapping of the following three types: (i) additive isotropic Gaussian noise; (ii) masking noise where a fraction of the elements of x (randomly chosen) is forced to set to 0; (iii) salt-and-pepper noise in which a fraction of the elements of x (chosen at random) is set to its minimum or maximum value (typically, 0 or 1).
- Second, mapping the corrupted input $\tilde{x}$ to a hidden representation y, $y = f_\theta(\tilde{x})$ and successively from which a reconstruction z is yielded by another mapping, $z = g_{\theta'}(y)$.

- Third, minimizing the average reconstruction error between the uncorrupted input x and z reconstructed from the corrupted $\tilde{x}$ over a training set by the stochastic gradient descent algorithm.

Stacked denoising autoencoders can be used similarly with stacked RBMs in DBNs to initialize a deep network [2, 14]. A special characteristic of stacked denoising autoencoders is that the input corruption is only used for the initial denoising — training of each individual layer to learn useful feature extractors. Once the mapping f_θ is learned, it will then be used on uncorrupted inputs to produce the representation, hence a clean input for training the next layer. The complete procedure for learning and stacking several layers of denoising autoencoders is shown in Fig. 9.10.

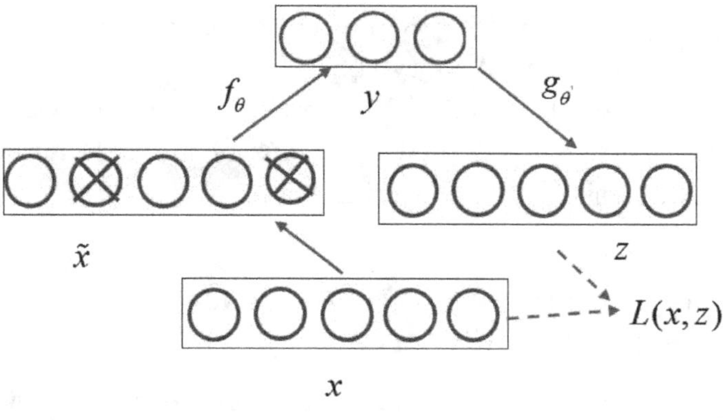

Fig. 9.9. The framework of a denoising autoencoder.

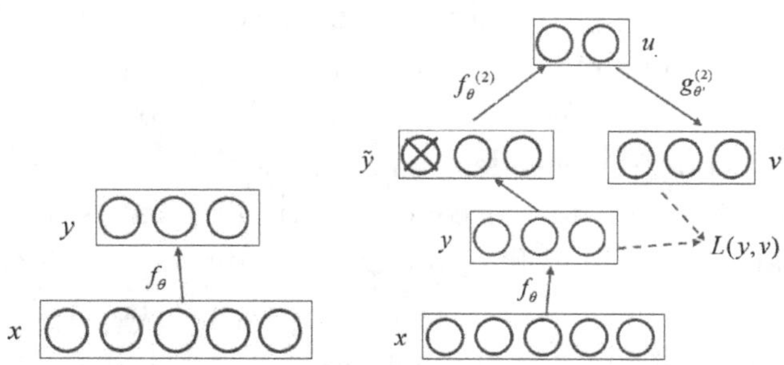

Fig. 9.10. The training process of a stacked denoising autoencoder.

The weights learned from the corrupted data are more robust than those from clean training data. On the other hand, corrupted data can alleviate the gap between training data and testing data.

A novel feature learning method based on sparse autoencoders was proposed to learn depth features for human body detection [47]. The automatically learned features can capture the intrinsic human body structure, which greatly benefits human body detection. Short text vectors are usually non-uniform, long, high dimensional and sparse, which are difficult to cluster using conventional clustering algorithms [24]. A feature extraction and clustering algorithm based on a deep autoencoder was proposed in the study, where high-dimensional text vectors were transformed into essentially low-dimensional feature vectors by a sparse denoising autoencoder [25]. Experimental result showed that the extracted text features can significantly improve clustering performance. Autoencoders combined with Long–Short Term Memory (LSTM) were used to cluster short text vectors in another study [18]. Moreover, a missing value imputation algorithm was proposed by means of a sparse autoencoder combined with fast-search clustering [3].

9.3.5. *Applications of deep autoencoders in pattern recognition*

On the basis of the minimum description length principle, an objective function was derived for training autoencoders to discover nonlinear factorial representations [16]. A deep autoencoder network pretrained by restricted belief machines was proposed for data dimensionality reduction [15]. A multilayer autoencoder with a small central layer can convert high-dimensional data to low-dimensional codes in encoders and reconstruct the coder to high-dimensional outputs in a decoder. Pretraining is used to provide good initial weights, and fine-tuning can improve the global performance of autoencoders. Both techniques ensure the success of deep autoencoders trained using gradient descent algorithms. As a nonlinear generalization of PCA, the experimental results demonstrate that a multilayer autoencoder performs more efficiently than PCA and logistic PCA.

An original strategy for building deep network is based on denoising autoencoders which is pretrained locally using corrupted inputs [50]. The prominent contribution of the denoising criterion as a tractable unsupervised objective can extract useful higher level representations. The experimental results indicate that a denoising autoencoder provides significantly lower classification errors that are almost as low as those provided by DBNs. Denoising autoencoders can learn Gabor-like edge detectors and larger stroke detectors, compared with ordinary autoencoders.

To reduce the risk of overfitting of stacked autoencoders, a modified stacked autoencoder based on the dynamic dropout strategy was proposed for spam classification, namely dd-SAE [45]. The Matthews correlation coefficients of the dd-SAE outperform those of stacked autoencoders on a benchmark of error datasets. In addition, a modified denoising autoencoder with clustering was used to supply the missing value [3].

Sparsity was first introduced in the context of sparse coding in the visual system in computational neuroscience [33]. Sparsity has become a key element in deep convolutional networks and DBNs, taking advantage of a variant of autoencoders with a sparse distributed representation. Several state-of-the-art related autoencoders are reviewed briefly in previous sections. Alternatively, rectified nonlinearity activation [44] was introduced into neural networks, which leads to real zero activations instead of a sparsity penalty taking small but non-zero activation or firing probability [8]. A rectified linear unit (ReLu $f(x) = \max(0, x)$) is a one-sided saturated function that must take the zero value and thus implies a truly sparse representation, whereas the softplus function ($f(x) = \log(1 + e^x)$) [6] can be viewed as a smooth version of the ReLu function. Despite taking advantage of semi-supervised setups, deep rectifier networks can achieve the best performance even without unsupervised pretraining on supervised tasks. The experimental results obtained are close to the performance gap between neural networks learnt with and without unsupervised pretraining on image classification and text-based sentiment analysis tasks. Thus, the rectified neural network can be considered as a new milestone to understand the difficulty in training deep but purely supervised neural networks.

Both autoencoders and denoising autoencoders share the disadvantages of ignoring the 2D image structure that render them difficult when dealing with large images and introduce redundant parameters, forcing features to be global. Therefore, a fully connected convolutional autoencoder was proposed to overcome the limitation of ordinary autoencoders [29]. The convolutional autoencoders share the weights among locations in the input to preserve spatial locality in which stochastic gradient descent is used to update the weights. Followed by the technique, an extreme machine combined with a convolutional autoencoder was used to learn three-dimensional shape features to improve classification performance [53]. Alternative denoising autoencoders and contractive autoencoders were introduced by adding a well-chosen penalty term to the classical reconstruction cost function to yield robust features on the activation layer [37]. To overcome the high computational cost and scalability to high-dimensional features of stacked denoising autoencoders, a marginalized stacked denoising autoencoder was presented using linear denoisers, instead of random feature corruption, as the basic building blocks in the literature [4]. The improved approach using marginalized stacked denoising autoencoders significantly speeds up stacked denoising autoencoders by 2 orders of magnitude.

In parallel with research on unsupervised pretraining, such as stacked autoencoders, RBMs, and DBNs, convolutional networks have been developed and used successfully to advance state-of-the-art discriminative classification. Discriminative deep learning methods usually outperform these unsupervised pretraining techniques despite the latter providing better deep representations in an unsupervised manner. Considering that convolutional networks are well-suited for data, resulting in higher classification accuracy, convolutional autoencoders were proposed [27]. Convolutional autoencoders incorporate convolutional architectures within an autoencoder framework. Likewise, a winner-take-all method for learning hierarchical sparse representations in an unsupervised manner was proposed by Makhzani and Frey [28]. A fully connected winner-take-all autoencoder was initially explored, in which mini-batch statistics were used to directly enforce lifetime sparsity in the activations of the hidden units. The convolutional winner-take-all autoencoder for learning shift-invariant sparse representations was then proposed. By using such an autoencoder, spatial sparsity within each feature map, in addition to lifetime sparsity [32], was achieved using winner-take-all activation functions. The network is trained layer by layer, with spatial sparsity achieved in each layer by using the winner-take-all activation function within each feature map. Winner-take-all is a computing principle by which neurons in a layer compete with one another, and only the neuron with the highest activation is active while all other neurons are inactive [9, 34]. The experimental results demonstrated that winner-take-all autoencoders can be used to learn deep sparse representations and achieve competitive classification performance.

Compared with layer-by-layer autoencoder pretraining, alternative novel techniques, such as batch normalization used to improve initialization, can easily lead to enhanced performance [19]. A residual network (ResNet) becomes the other popular technique used to train very deep networks after wining the first place in the ILSVRC2015 classification task [10]. An identity function is subtly added into deep CNN to reformulate ResNet, which perform surprisingly with the increase of the hidden layers. However, some researchers considered very deep ResNet as an ensemble of multiple subnets, whereas most of the subsets are shallow nets [48].

9.4. Conclusion

It is well known that deeper BP networks can provide better classification/recognition performance at the cost of highly difficult training. Better initialization of networks is a necessary way to alleviate the gradient diffusion phenomenon in the process of error backpropagation of the BP algorithm. On the other hand, autoencoders are a type of unsupervised learning schemes, which attempt to render the outputs approximately equal to the inputs. The weights of deep networks determined by autoencoders are the better initialization that easily lead to the success of the training process. To improve the robustness and generalized capability, sparsity and noisy data are two core factors introduced into autoencoders, namely sparse autoencoders and denoising autoencoders. A comprehensive review of the state-of-the-art literatures about autoencoders in pattern recognition field is addressed in the end of this study. Moreover, deep autoencoders mainly focus on data denoising, dimensionality reduction and new data production. This study hence offers an overview of autoencoders.

Acknowledgments

This work is supported by Natural Science Foundation of Hebei Province (F2016201161), National Natural Science Foundation of China (61473111), and the special fund for one university in one province.

Bibliography

[1] Bengio, Y. Learning deep architectures for AI. *Found. Trends Mach. Learn.* **2**(1), 1–127, 2009.

[2] Bengio, Y., Lamblin, P., Popovici, D. and Larochelle, H. Greedy layer-wise training of deep networks. In *Advances in Neural Information Processing Systems* (*NIPS'06*), 153–160, 2006.

[3] Bu, F., Chen, Z. and Zhang, Q. Missing value imputation algorithm based on clustering and autoencoder. *Comput. Eng. Appl.* **51**(18), 13–17, 2015 (in Chinese).

[4] Chen, M., Xu, Z. E., Weinberger, K. Q. and Sha, F. Marginalized denoising autoencoders for domain adaptation. In *The 29th International Conference on Machine Learning* (*ICML'12*), Vol. 1206, pp. 46–83, 2012.

[5] Cho, Y. and Saul, L. K. Kernel methods for deep learning. In *Advances in Neural Information Processing Systems* (*NIPS'09*), Vol. 22, pp. 342–350, 2010.

[6] Dugas, C., Bengio, Y., Belisle, F. and Nadeau, C. Incorporating second-order functional knowledge for better option pricing. In *Advances in Neural Information Processing Systems (NIPS'01)*, 2001.

[7] Erhan, D., Bengio, Y., Courville, A., Manzagol, P.-A., Vincent, P. and Bengio, S. Why does unsupervised pre-training help deep learning? *J. Mach. Learn. Res.* **11**, 625–660, 2010.

[8] Glorot, X., Bordes, A. and Bengio, Y. Deep sparse rectifier neural networks. In *The 14th International Conference on Artificial Intelligence and Statistics (AISTATS)*, pp. 315–323, 2011.

[9] Grossberg, S. Contour enhancement, short-term memory, and constancies in reverberating neural networks. *Stud. Appl. Math.* **52**(3), 213–257, 1973.

[10] He, K., Zhang, X., Ren, S. and Sun, J. Deep residual learning for image recognition. In *The IEEE Conference on Computer Vision and Pattern Recognition (CVPR'16)*, pp. 1–9, 2016.

[11] Hebb, D. O. *The Organization of Behavior: A Neuropsychological Theory.* John Wiley & Sons, New York, 1949.

[12] Hinton, G., Srivastava, N., Krizhevsky, A., Sutskever, I. and Salakhutdinov, R. R. Improving neural networks by preventing co-adaptation of feature detectors. *Comput. Sci.* **3**(4), 212–223, 2012.

[13] Hinton, G. E. Learning multiple layers of representation. *Trends Cogn. Sci.* **11**(10), 428–434, 2007.

[14] Hinton, G. E., Osindero, S. and Teh, Y.-W. A fast learning algorithm for deep belief nets. *Neural Comput.* **18**, 1527–1554, 2006.

[15] Hinton, G. E. and Salakhutdinov, R. R. Reducing the dimensionality of data with neural networks. *Science* **313**, 504–507, 2006.

[16] Hinton, G. E. and Zemel, R. S. Autoencoders, minimum description length and Helmholtz free energy. In *Advances in Neural Information Processing Systems (NIPS'93)*, pp. 1–9, 1994.

[17] Hornik, K. Multilayer feedforward networks are universal approximators. *Neural Netw.* **2**, 359–366, 1989.

[18] Huang, J. and Deng, M. A short text clustering method based on lstm-autoencoder. *Comput. Technol. Autom.* **3**, 75–80, 2017 (in Chinese).

[19] Ioffe, S. and Szegedy, C. Batch normalization: Accelerating deep network training by reducing internal covariate shift. In *The 32nd International Conference on Machine Learning (ICML'15)*, Vol. 37, pp. 1–9, 2015.

[20] Krizhevsky, A., Sutskever, I. and Hinton, G. E. Imagenet classification with deep convolutional neural networks. In *Advances in Neural Information Processing System (NIPS'12)*, pp. 1097–1105, 2012.

[21] LeCun, Y., Bengio, Y. and Hinton, G. Deep learning. *Nature* **521**, 436–444, 2015.

[22] LeCun, Y., Boser, B., Denker, J. S., Henderson, D., Howard, R. E., Hubbard, W. and Jackel, L. D. Backpropagation applied to handwritten zip code recognition. *Neural Comput.* **1**(4), 541–551, 1989.

[23] Lin, S.-H., Kung, S.-Y. and Lin, L.-J. Face recognition/detection by probabilistic decision-based neural network. *IEEE Trans. Neural Netw.* **8**(1), 114–132, 1997.

[24] Liou, C.-Y., Cheng, W.-C., Liou, J.-W. and Liou, D.-R. Autoencoder for words. *Neurocomput.* **139**, 84–96, 2014.

[25] Liu, K. and Yuan, Y. Short texts feature extraction and clustering based on auto-encoder. *Acta Sci. Natur. Univ. Pekinensis* **51**(2), 282–288, 2015 (in Chinese).

[26] Makhzani, A. and Frey, B. k-sparse autoencoders. In *International Conference on Learning Representations (ICLR'14)*, pp. 1–9, 2014.

[27] Makhzani, A. and Frey, B. A winner-take-all method for training sparse convolutional autoencoders. Preprint, arXiv:1511.06247v3, pp. 1–9, 2014.

[28] Makhzani, A. and Frey, B. Winner-take-all autoencoders. In *Advances in Neural Information Processing Systems (NIPS'15)*, pp. 2791–2799, 2015.

[29] Masci, J., Meier, U., Ciresan, D. and Schmidhuber, J. Stacked convolutional auto-encoders for hierarchical feature extraction. In *International Conference on Artificial Neural Networks*, Lecture Notes in Computer Science, Vol. 6791, pp. 52–59, 2011.

[30] McCulloch, W. S. and Pitts, W. A logical calculus of the ideas immanent in nervous activity. *Bull. Math. Biophys.* **5**(4), 115–133, 1943.

[31] Minsky, M. and Papert, S. *Perceptrons: An Introduction to Computational Geometry.* The MIT Press, Cambridge, MA, 1969.

[32] Ngiam, J., Koh, P., Chen, Z., Bhaskar, S. and Ng, A. Y. Sparse filtering. In *Advances in Neural Information Processing Systems (NIPS'11)*, pp. 1–9, 2011.

[33] Olshausen, B. A. and Field, D. J. Sparse coding with an overcomplete basis set- a strategy employed by v1. *Vision Res.* **37**, 3311–3325, 1997.

[34] Oster, M., Douglas, R. and Liu, S. Computation with spikes in a winner-take-all network. *Neural Comput.* **21**(9), 2437–2465, 2009.

[35] Phung, S. L. and Bouzerdoum, A. A pyramidal neural network for visual pattern recognition. *IEEE Trans. Neural Netw.* **18**(2), 329–343, 2007.

[36] Ranzato, M., Boureau, Y.-L. and LeCun, Y. Sparse feature learning for deep belief networks. In *Advances in Neural Information Processing Systems (NIPS'07)*, Vol. 20, pp. 1185–1192, 2007.

[37] Rifai, S., Vincent, P., Muller, X., Glorot, X. and Bengio, Y. Contractive auto-encoders: Explicit invariance during feature extraction. In *The 28th International Conference on Machine Learning (ICML'11)*, 2011.

[38] Rosenblatt, F. The perceptron: A probabilistic model for information storage and organization in the brain. *Psychol. Rev.* **65**(6), 386–408, 1958.

[39] Rosenblatt, F. *Principles of Neurodynamics: Perceptrons and the Theory of Brain Mechanisms.* Spartan Books, Washington, DC, 1962.

[40] Rowley, H. A., Baluja, S. and Kanade, T. Neural network-based face detection. In *The IEEE Conference on Computer Vision and Pattern Recognition (CVPR'96)*, pp. 203–208, 1996.

[41] Rowley, H. A., Baluja, S. and Kanade, T. Neural network-based face detection. *IEEE Trans. Pattern Anal. Mach. Intell.* **20**(1), 23–38, 1998.

[42] Rumelhart, D. E., Hinton, G. E. and Williams, R. J. *Learning Internal Representations by Error Propagation*, Vol. 1, pp. 318–362. MIT Press, Cambridge, MA, 1986.

[43] Rumelhart, D. E., Hinton, G. E. and Williams, R. J. Learning representations by back-propagating errors. *Nature* **323**(6088), 533–536, 1986.

[44] Salinas, E. and Abbott, L. F. A model of multiplicative neural responses in parietal cortex. *Neurobiology* **93**(21), 11956–11961, 1996.

[45] Shen, C., He, J. and Deng, Y. Spam filtering based on modified stack autoencoder. *J. Comput. Appl.* **36**(1), 158–162, 2016 (in Chinese).

[46] Simonyan, K. and Zisserman, A. Very deep convolutional networks for large-scale image recognition. Preprint, arXiv:1409.1556, 2014.

[47] Su, S.-Z., Liu, Z.-H., Xu, S.-P., Li, S.-Z. and Jia, R. Sparse autoencoder based feature learning for human body detection in depth image. *Signal Process.* **112**, 43–52, 2015.

[48] Veit, A., Wilber, M. and Belongie, S. Residual networks are exponential ensembles of relatively shallow networks. In *Advances in Neural Information Processing Systems (NIPS'16)*, pp. 1–9, 2016.

[49] Vincent, P., Larochelle, H., Bengio, Y. and Manzagol, P.-A. Extracting and composing robust features with denoising autoencoders. In *The 25th International Conference on Machine Learning (ICML'08)*, pp. 1096–1103, 2008.

[50] Vincent, P., Larochelle, H., Lajoie, I., Bengio, Y. and Manzagol, P.-A. Stacked denoising autoencoders: Learning useful representations in a deep network with a local denoising criterion. *J. Mach. Learn. Res.* **11**, 3371–3408, 2010.

[51] Werbos, P. *Beyond Regression: New Tools for Prediction and Analysis in the Behavioral Sciences*, Ph.D. thesis, Harvard University, 1974.

[52] Weston, J., Ratle, F. and Collobert, R. Deep learning via semi-supervised embedding. In *The 25th International Conference on Machine Learning (ICML'08)*, pp. 1168–1175, 2008.

[53] Xie, Z., Wang, Y., Dou, Y. and Xiong, Y. 3D feature learning via convolutional auto-encoder extreme learning machine. *J. Comput. Aided Des. Comput. Graph.* **27**(11), 2058–2064, 2015 (in Chinese).

Chapter 10

Genetic Algorithm: From Promiscuity to Monogamy

Ting Yee Lim[*] and Choo Jun Tan[†]

School of Science and Technology
Wawasan Open University
54, Jalan Sultan Ahmad Shah, Gerge Town
10050 Penang, Malaysia
[*]*tylim1@wou.edu.my*
[†]*cjtan@wou.edu.my*

Genetic algorithm (GA) is one of the most widely used evolutionary algorithms to date. Inspired by the natural evolution, GA operates on a population of candidate solutions that are transformed into potential solution(s) over time via different variation operators including selection, crossover and mutation. Breaking the convention, this paper describes a variant of GA that takes on the concept of social monogamy. Monogamous parents are allowed to breed over several generations before a reselection of new pair is formed. Sporadically, infidelity occurs to induce diversity, disperse genetic information across the population and expedite convergence. The motivation and conceptualization of monogamous GA is discussed thoroughly in this paper. It shall be evidential that monogamy is indeed possible within GA, which initially held high within promiscuity. As biologists struggle to gain insights into the presence of promiscuity against monogamy, and vice versa, perhaps a computing model, in particular the evolutionary algorithm based experiments can shed some lights.

10.1. Introduction

Genetic Algorithm (GA) [10] is a class of the Evolutionary Algorithm (EA). Borrowing the idea of natural selection and learning process, EA establishes an effective computing system for problem solving [22]. The search progresses in parallel by maintaining a pool of candidate solutions, each known as an *individual* or *chromosome* in the context of GA. In turn, an individual is associated with a *fitness value*, which is evaluated through a problem-specific *objective function*. The fitness value determines the quality of the individual. Fitter individuals are often preferred by GA. The population evolves (iterates) through repeated applications of various operations, such as selection, mutation and crossover until some predefined termination conditions are satisfied. Each iteration step is referred to as a *generation*.

With the proliferation of studies in GA over the years, a plethora of GA variants [11, 12, 23] has been discussed in the literature. A common thread shared by these variants is that individuals are breeding promiscuously. In other words, once a pair of parents are selected for breeding, there is no guarantee that they will be paired up again in the next generation. In generational GA [7], offspring typically replace their immediate parents. Therefore, instantly eliminating the chances for the same pair to rebreed.

On the contrary, Nature continuously exhibits the strength of monogamous mating system. In its simplest form, monogamy is a practice or state whereby an individual forms an enduring relationship with only one partner at any one time or throughout his or her lifetime. It is particularly common in birds, but also occurrs in fish, rodents, lizards and modern human civilization.

If the Nature has once inspired GA, could monogamy repeat the success or even improve the optimization ability of GA? In order to answer these questions, this chapter takes you through the routes of identifying the strengths of monogamy in natural population (see Section 10.2), and why it is possible for GA to learn from Nature once again (Section 10.3.1). Inherently, a new variant of GA, namely the Monogamous Pairs Genetic Algorithm (MopGA) [15] is discussed in Section 10.3.2, followed by its enhanced version in Section 10.3.3.

In MopGA, parents are bonded and bred for several consecutive generations, after which new pairs of parents are selected from the population. Offspring produced by monogamous parents go through sibling competition to ensure only the best are qualified for the next generation. Sporadically, infidelity occurs to induce diversity, disperse genetic information across the population and expedite convergence. The MopGA presents several interesting behaviour worth commenting. This can be found in Section 10.4. Lastly, a summary of the chapter is given in Section 10.5.

10.2. Monogamy in Natural Population

In a monogamous society, mating partners display selective but non-exclusive affiliation and biparental care of offspring. As opposed to polygyny, polyandry or polyamory, an individual is attached to only one mating partner at any one time. Pair bond is a form of intense social attachments.

Interestingly, 90% of birds are socially monogamous [20]. Researchers [8] have found that the duration of pair bond is a determinant of the breeding attempts, hatching and fledging in socially monogamous bearded reedling (*Panrus biarmicus*).[a] High success rates are predicted with longer pair bond duration. Some of the highly acclaimed benefits of monogamy include assurance of partner's reproductive capacity, care for the well being of offspring and longer live span for monogamous parents themselves.

More recently, Resendez and Aragona [18] suggest that monogamous pair bonds are not merely coincidental occurrences but regulated by neural mechanisms. Under the influence of endogenous opioid transmission, positive stimuli are recognized as reward,

[a]Also known as bearded parrotbill or bearded tit, this bird is small in size, orange–brown in colour and has a long tail.

whereas aversion if harmful. Likewise, bond is formed when the parties involved acknowledge that the neural event is indeed rewarding. Subsequently, these individuals attempt to maintain the relationship by showing preference to only one mate and rejecting others. During the absence of a partner, the stress and negative effect induced will further promote pair bond maintenance.

Surprisingly, infidelity or extrapair is not uncommon especially among socially monogamous birds [9]. A new finding reports that birds are more inclined to infidelity under unpredictable climatic conditions [1]. This could be explained using the good genes and future partnerships hypotheses [17]. Seeking out diverse genes for their offspring in an uncertain environment is likely to improve their chances of survival. Meanwhile, the latter provides opportunity for more favourable mates in the future.

10.3. Artificial Monogamy

In this section, we uncover why and how monogamy is incorporated into GA.

10.3.1. *Why monogamy in GA?*

The major motivation is led by the acknowledgement that crossover is the main genetic operator in GA [7, 10]. Crossover is also sometimes regarded as the exploitation operator [4]. As a result, monogamy can be introduced to improve the exploitation of the search space, in particular between monogamous parents.

Moreover, as with the unpredictable climatic conditions of natural evolution, GA is a heuristic search method. By heuristic, if one is lucky, one may find a good or even an optimal solution to the problem. Unfortunately when luck is not around the corner, one will not find any solution or even if there is, it may be far from the optimal. Similarly, the standard crossover operation does not always produce better fit offspring relative to their parents. The quality of the offspring is a factor of chance with every application of the crossover. Hence, by repetitive application of crossover operations, the chances of discovering a better fit offspring increases. At the same time, a more thorough exploitation of the current search region can be achieved.

Now in order to appreciate the next motivation, it would be wise to comprehend the term *exploratory power* of a crossover operation. The exploratory power of a crossover operator simply refers to the manner in which the operator alters individuals within the subspaces of each dimension or hyperplane [21]. Exploratory power is particularly useful when the population size is small and at early stages of the evolution so that a wide search space is well sampled. The more points involved in a crossover, the higher its exploratory power. For example, the uniform crossover may generate an offspring anywhere in between its subspaces, whereas a 1-point crossover generates an offspring limited to only a small subset of the search space. Nevertheless, Ref. [5] shows that the less disruptive crossover operators like 1- and 2-point crossover can also achieve higher exploratory power like the uniform crossover through repetitive applications of crossover operation over successive generations. In this light, monogamy can facilitate a simple crossover operation in achieving higher exploratory power.

To give a better picture of how monogamy works for GA, let us consider the following situations: Suppose that we apply a 1-point crossover on individuals from two hyperplanes (shown as Parent 1 and Parent 2, respectively, in Table 10.1). "#" denotes the symbol common to both individuals. As can be seen from the above examples, the closer the parents are in terms of genotype distance (the higher the numbers of "#"), the less likely the number of unique offspring produced by the 1-point crossover.

Generally, the number of unique offspring, U may be formulated as [13]

$$U = 2\left[(L-1) - \sum \#\right] \qquad (10.1)$$

where L represents the chromosome length. In Eq. (10.1), the multiplication by 2 is due to the assumption that every crossover produces two offspring. Broadly speaking, it is possible to produce

Table 10.1. Sample outputs of 1-point crossover on two individuals from different hyperplanes.

No.	Parent 1	Parent 2	Offspring	Number of unique offspring, U
1.	0000	1111	0111, 0011, 0001, 1000, 1100, 1110	6
2.	0#10	1#01	0#01, 1#10, 0#11, 1#00	4
3.	0##0	1##1	0##1, 1##0	2
4.	0#1#	1#0#	0#0#, 1#1#	2
5.	0###	1###	0###, 1### (*similar to parents*)	0

more unique offspring from parents that are genotypically dissimilar or have longer chromosome length. For example, consider when $L = 200$, $\sum \# = 20$; hence 358 different offspring may be generated according to Eq. (10.1).

Nonetheless, worth noting that these offspring may fall inside or outside their parents' hyperplanes. We refer to the event as a *disruption* when the offspring generated fall outside the defining boundaries of their parents (i.e. breaking a sample lower order schema). Disruption is required in the exploration of a new search space. However, the usefulness of exploration is subjected to the objective function such that the search process produces feasible solutions. Importantly, it should be moving towards promising regions. The latter can be accomplished via a selection mechanism that biases fit individuals.

In sum, these examples reinforce the idea that exploratory power of crossover operations can be achieved via monogamy, in which the same pair of parents mate multiple times, increasing their chances of discovering better solutions.

10.3.2. *Monogamous pairs genetic algorithm*

With the theoretical background and motivations laid, the realization of artificial monogamy is finally presented in the form of monogamous pairs genetic algorithm (MopGA).

A comprehensive description of the generic MopGA framework is given in [15]. Generally, the algorithm (Algorithm 10.1) works as follows: The evolutionary process begins by randomly initializing

Algorithm 10.1 MopGA

```
1: Initialize N number of individuals and randomly pair them to form N/2 families.
2: while termination condition not reached do
3:     for each family do
4:         if g_p is due then
5:             Perform rebond
6:         else
7:             if P_i is satisfied then
8:                 Perform infidelity
9:             end if
10:        end if
11:        Perform breeding (crossover w.p P_c = 1; mutation w.p P_m)
12:        Evaluate fitness of offspring
13:        Offspring competition
14:    end for
15: end while
```

Note: w.p stands for "with probability".

N numbers of individuals and pairing them up as monogamous parents (line 1). A pair of monogamous parents and their product of reproduction form a family. Hence, there are $\frac{N}{2}$ numbers of families within a MopGA population.

At each generation, each pair of monogamous parents goes through the breeding cycle as long as their bond session is still valid. The length of a bond is controlled by the *pair bond tenure*, g_p parameter. For instance, if $g_p = 100$, then each monogamous parents pair will breed for 100 consecutive generations. After which, the bond is broken (*rebond*), and selection followed by formation of new monogamous parents will take place (line 5).

Moreover, as with natural monogamy, *infidelity* may proceed at probability P_i (also known as infidelity rate) during a pair bond session (line 8). In other words, instead of breeding with the current partner, one of the parents (usually the better fit parent) will mate with a randomly selected individual (known as extrapair) from the population. Infidelity requires that the extrapair be any individual other than from the current family. The simple incest prevention strategy is proposed to maintain a higher level of population diversity. Furthermore, the notion of infidelity is to encourage exploration of the search space and alleviate premature convergence, while also speeding up convergence (the assertion has been proven empirically elsewhere [14]).

Breeding is the main reproduction process, which encompasses both the crossover and mutation operations (line 11). It is worth remarking that the crossover operation is mandatory in MopGA; that is, crossover rate, P_c is always 1.0, while mutation will occur at probability rate P_m. Unlike conventional GAs, it is unnecessary to explicitly reserve unaltered individuals for the next generation since MopGA preserves individuals in the form of monogamous pairs. To put it simply, "old" population (represented by monogamous parents) and "new" population (represented by the offspring) coexist in every generation. Meanwhile, as only two offspring will be preserved in each family, existing offspring and the newly generated offspring contest against each other for survival into the next generation. This phenomenon is coined as *offspring competition* (line 13). Implicitly, elitism is achieved — the best individuals are always preserved automatically. According to Rudolph [19], GA will converge asymptotically to a global optimum in the presence of elitism.

At rebond, the best offspring from each family will automatically serve as the parent of a new family. Its mating partner will then be selected randomly from the remaining population guided by the incest prevention technique explained in the preceding paragraph (line 5). Finally, the evolutionary process repeats until all stopping conditions are satisfied.

10.3.3. *Adaptive monogamous pairs genetic algorithm*

The MopGA in the previous section introduces two extra parameters to GA, namely g_p and P_i. Parameter tuning may appear as a hindrance to the effective use of an algorithm. Hence, the authors [13] proposed an enhancement to the original MopGA. It is coined as the adaptive monogamous pairs genetic algorithm (AMopGA).

Both monogamous GAs share the same framework as described in Algorithm 10.1, with further enhancement introduced in AMopGA as follows:

- Adaptive parameters:
  - Mutation rate (line 11);
  - Pair bond tenure (line 5);
  - Infidelity rate (line 8).
- New features:
  - Perform infidelity with secondary mutation on elite individual (line 8);
  - Multiple adaptive crossover operators (line 11).

The authors cautioned that the new features (secondary mutation and adaptive crossover operators) are rather problem-specific. They are optional tools that could be implemented on a case by case basis. On the bright side, these tools can be adopted into the monogamy framework with little effort. Their effects will be studied in Section 10.4. Due to space limitation, we leave the detailed implementation of AMopGA for interested readers who can refer to [13].

10.4. Does Monogamy Really Work for GA?

This section presents the preliminary empirical investigation into the impact of monogamy on GA's evolutionary process. The notion is to gain better insights at how monogamy affects the performance of GA for different configuration choices.

In order to achieve the objectives, the royal road (RR) function is adopted. It has been used extensively in the study of GA behaviour [6, 16]. The objective of RR is to maximize discrete building blocks, i.e. short, low-order schemata. The RR can be formulated as in Eq. (10.2) [24].

$$f_{rr}(B) = \sum_{i=1}^{M} g(b_i) \tag{10.2}$$

$$g(b) = \begin{cases} K, & \text{if } u(b) = K \\ 0, & \text{otherwise} \end{cases} \tag{10.3}$$

where B is a block represented by $\{x_1, x_2, \ldots, x_L\}$ bits and L denotes the chromosome length with $L = MK$. Additionally, the ith sub-block of B is given by $b_i = \{x_{i(1)}, x_{i(2)}, \ldots, x_{i(K)}\}$. Note that the sub-function g defines the fitness of the individual block according to Eq. (10.2), where $u(b)$ is the unitation of b. That is, when all bits in a sub-block are ones, g returns the value K, otherwise zero. In a typical RR function, both M and K typically carry the values 8 with the known optimal fitness of 64.

All experiments in this section are conducted on population with size $N = 500$, 2-point crossover and bit-flip mutation at $P_i = \frac{1}{L}$. The maximum iteration G is set to 2000, $P_i = 0.6$ and $g_p = 0.02G$. Furthermore, for comparison purposes, the standard GA (SGA) with the following parameters are used as a control algorithm: $P_c = 0.85$, $P_m = \frac{10}{L}$, roulette wheel selection with elitism (10% of the best individuals will be included in the subsequent generation). The remaining parameters are the same as MopGA. Each experiment runs independently for 50 trials. Note that pair bond rate and infidelity rate are adaptive in AMopGA, hence no setting is required. Furthermore, in AMopGA, two additional crossover operators are employed adaptively, namely 1-point crossover and uniform crossover. Also, t denotes the generation number.

10.4.1. *Solution quality*

First and foremost, Fig. 10.1 summarizes the comparative results of AMopGA, MopGA and control algorithm SGA in solving the RR problem. While all three algorithms chalk a success rate of 100%, the AMopGA achieves this in the least mean number of generations,

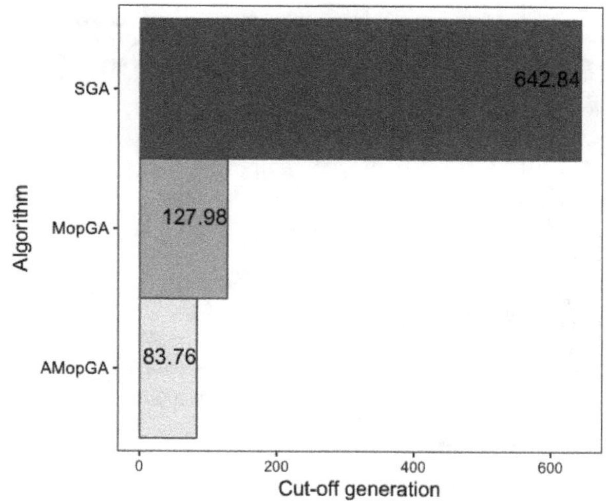

Fig. 10.1. Comparing the performance of AMopGA, MopGA and SGA in solving the RR problem.

denoted by the cut-off generation.[b] In comparison to MopGA, this leads to a saving of 34.55% in the cut-off generation.

10.4.2. *Convergence behaviour*

10.4.2.1. *Monogamous pairs genetic algorithm*

In order to comprehend the convergence behaviour of monogamous GAs, the evolution of population best fitness is plotted in Fig. 10.2. Interestingly, the MopGA produces a "ladder"-like curve with the width of each ladder (step) corresponding to a pair bond tenure length, i.e. $g_p = 40$ generations (0.02×2000). Reproduction and sibling competition occur in isolation guarded by the family structure within each ladder rung.

[b]Cut-off generation is defined as the number of generations in which the search algorithm reaches the desired precision. The sum of all cut-off generations over the total number of trial run gives the mean cut-off generation, $= \frac{1}{T}\sum_{r=1}^{T} g_r$, where T represents the number of trials and g_r gives the cut-off generation at the rth trial. An efficient algorithm generally has a smaller mean cut-off generation.

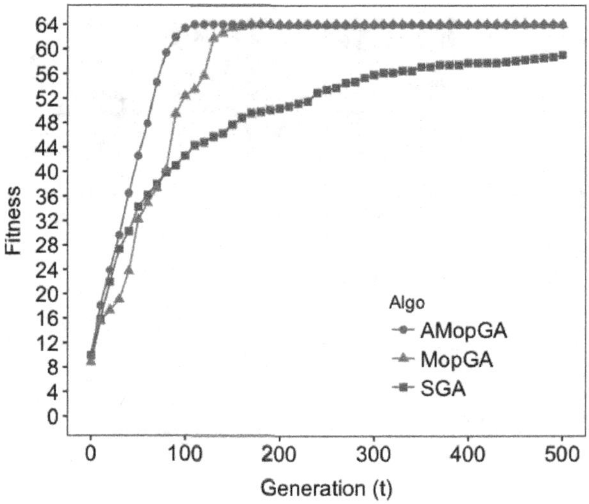

Fig. 10.2. Convergence plots sampled at every 10th generation.

At the end of pair bond tenure, rebond takes place and new families are formed through a selection mechanism. At this time, the genetic materials are shared among all members of the entire population. Priority is given to the best offspring to automatically serve as the first parent of a new family, while randomly selecting the second parent from the remaining individuals in the pool. The main notion behind this is to make the reproduction process generate offspring of better fitness. As crossover is performed between parents that differ genotypically or phenotypically, and one is highly fit, the product is likely an offspring of intermediate to high fitness value. Consequently, the population experiences improvement in fitness in the form of "climbing ladder".

10.4.2.2. *Adaptive monogamous pairs genetic algorithm*

At this point, we assume that the reader is familiar with the concept of AMopGA as contained in [13].

The appearance of stepwise increment in fitness is less obvious in AMopGA. A smoother convergence curve is observed in AMopGA, similar to that of SGA. The reason can be explained considering family-wise adaptive parameters that occur at every point of the evolutionary process. Let us consider how this happens: In MopGA, pair bond tenure is controlled by a single global g_p value and every family experiences rebond at the same time. On the contrary, each family in AMopGA has full control over when to rebond and to continue the monogamous relationship depending on their performance and the duration of the bond. Generally speaking, the longer the bond the more likely is the occurrence of a rebond. On the other hand, the more effective the crossover operation, i.e. the ability to produce more better fit offspring, the less likely is the occurrence of a rebond.

Moreover, P_i varies in accordance with offspring survival rate, V. The greater the offspring survival (new offspring replacing old ones), the lower is the P_i, and vice versa. In order to give an idea of how the parameters change adaptively with the evolutionary environment, refer to Fig. 10.3. The representative plots depict a typical change in pair bond tenure rate, G_p and P_i was captured from a single family of AMopGA during the search process of solving a separate 64-bit Onemax[c] problem.

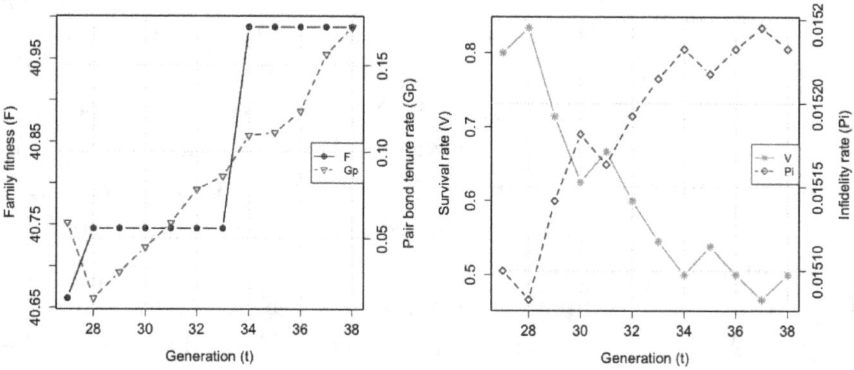

Adaptive G_p

(a)

Adaptive P_i

(b)

Fig. 10.3. Sample adaptive parameters in AMopGA when solving the Onemax problem.

[c]In the onemax problem, each solution is represented by zeros and ones. The objective of the problem is to find a solution with the maximum number of ones. Seemingly

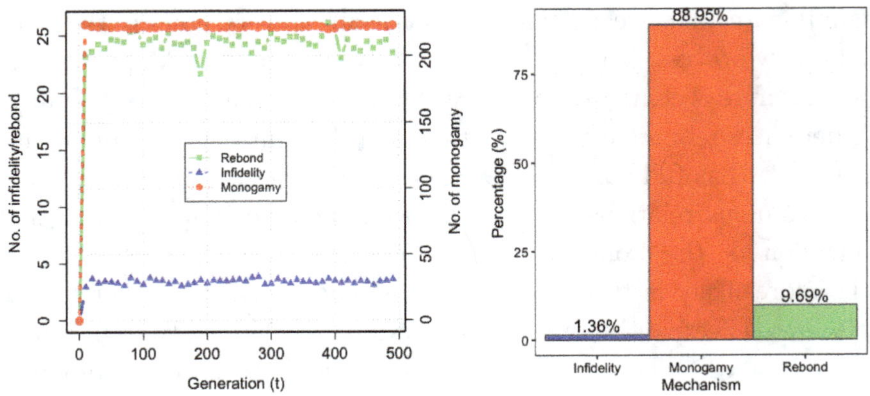

(a) (b)

Fig. 10.4. Distribution of infidelity, rebond and monogamy mechanisms in AMopGA.

Also note that AMopGA has the faster convergence rate compared to its predecessor and SGA. This can be attributed to the effective combined adaptation techniques employed. Figure 10.4(a) provides a close up view of the change in occurrence of monogamy, infidelity and rebond through the evolutionary process of AMopGA. Meanwhile, Fig. 10.4(b) informs us that monogamy, which is the hallmark of monogamous GA, encompasses 89.95% of the evolutionary process. Rebond takes up 9.69% of the search process in order to reconstruct a new family through the selection mechanism. Meanwhile, infidelity only occurs at the rate of 1.36%. Its task is to recombine monogamous parents with other extrapair individuals in order to boost the chances of discovering new potential offspring. As a quick comparison, suppose that MopGA parameters are $g_p = 0.02G$ and $P_i = 0.2$, then 2% and 20% of the total evolutionary time would be spent in rebond and infidelity mechanisms, respectively. The remainder of the time is spent on monogamous breeding.

---

simple, the problem actually possesses exponentially increasing number of local optima on landscape when considering a 1-point crossover operator [3].

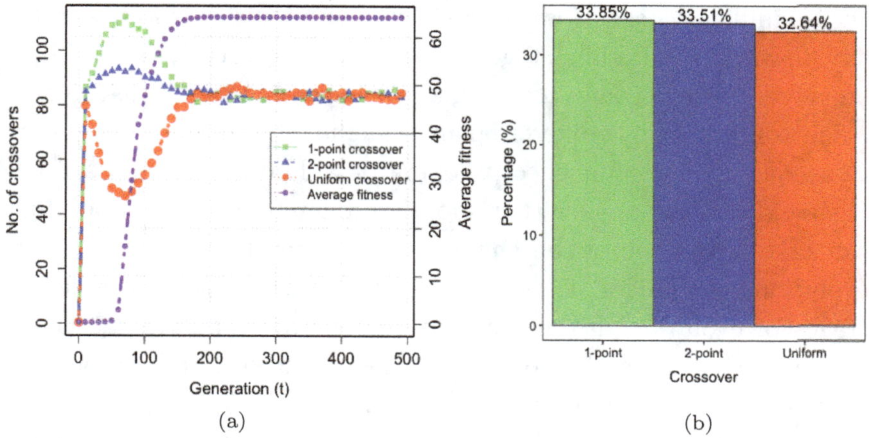

Fig. 10.5. Distribution of multiple and adaptive crossover operators in AMopGA.

Another significant difference between the two monogamous GAs is the presence of multiple and adaptive crossover operators in AMopGA. Let us illustrate this using Fig. 10.5. In general, there is almost equal distribution of usage among the three crossover operators. However, with further diagnosis, one finds fierce competition among these operators at the early stages of the evolutionary process, i.e. between 0 to roughly 200 generations (see Fig. 10.5(a)). Two key points worth commenting are as follows.

First, as the population converges (average population fitness line becomes horizontal), the crossover operator distributions usage also converge. The situation may not be as complex as first seen. Upon convergence, the population is generally populated by individuals who are closely related genotypically. Therefore, the selection of crossover operators become ineffective (similar to random operation) because no operator is better than the other in producing offspring.

Second, we emphasize again that diversity is essential for the adaptive selection of the crossover operator. For instance, at the beginning of the evolution, the 1- and 2-point crossovers are effective (refer to Fig. 10.5(a)). In comparison to the uniform crossover, the 1-point crossover works well in RR since long building blocks recombination is beneficial [2]. Uniform crossover tends to break these building blocks. However, as the population slowly looses diversity towards the middle stage of evolutionary process, the uniform crossover's role becomes more significant. The chances of discovering a new solution from a different plane is higher with uniform crossover. Consequently, the more effective operator will be selected more often. It should be clear by now that adopting different crossover operators can take advantage of the fact that the search process experiences evolutionary changes. Different operators work better at different stages of evolution. Employing various operators adaptively hence increases the probability of appointing the most suitable operator for circumstances unknown beforehand.

Finally, AMopGA also introduces secondary mutation at infidelity (Levy mutation is used in this case). To verify the efficacy of this feature, separate tests were made with and without secondary mutation. Further distinction is made between mutation on an individual with the best fitness (known as elite) and individual who is selected randomly. The violin plots in Fig. 10.6 inform us that secondary mutation on an elite individual is better than each of the other tests (with 0.05 significance level on t-tests). Nonetheless, it is worth commenting that the case may not be generalized and most likely problem-specific. The choice of the variant operator may also influence the results.

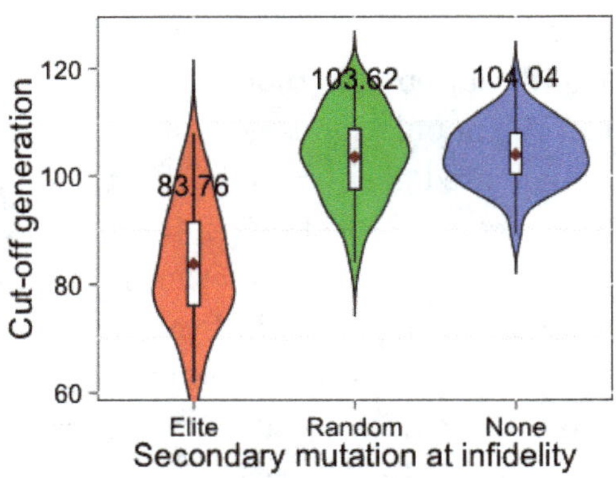

Fig. 10.6. Violin plots showing the density distribution of cut-off generation for AMopGA considering secondary mutation in infidelity. Red dots indicate the distributions' mean values.

10.4.3. *Diversity*

This section investigates the diversity maintenance ability of monogamous GAs. As revealed by Fig. 10.7, the AMopGA looses diversity much faster than both MopGA and SGA. The SGA has the highest population diversity.

As every coin has two sides, diversity is important to improve any search process since there will be more "routes" for the search to progress, and thus alleviating premature convergence. However, if diversity is too high, convergence becomes slow. Hence, when considering population diversity, one should take into account its usefulness. To put it simply, diversity is required to assist evolution into seeking better results. As the search hits the optimum points, diversity may be ignored (the search can be terminated). Similarly, while AMopGA has a lower diversity level compared to the others, diversity loss can be offset by the faster convergence rate to reach global optima. Recall that the AMopGA and MopGA found optimal solutions after about 84 and 128 generations, respectively (see Section 10.4.1). These are marked as vertical lines in Fig. 10.7. The SGA, meanwhile, only found an optimal solution after over 600 generations.

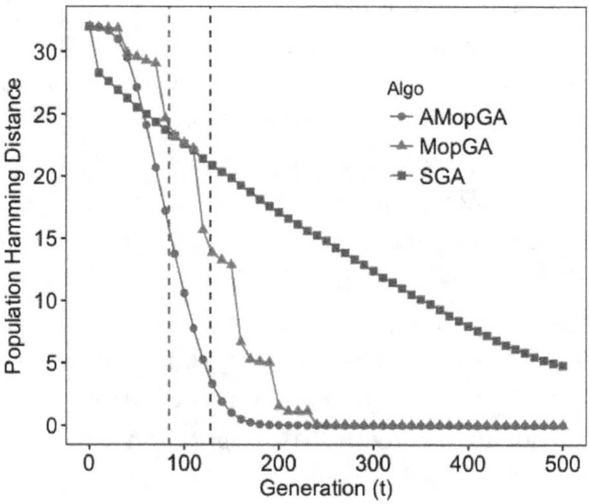

Fig. 10.7. Population diversity plot. Optimal solutions are found during generations marked by vertical lines.

Observe that MopGA once again displays stepwise changes, but this time in the population diversity. The end of each ladder rung is associated with the end of pair bond tenure. During the pair bond cycle, each family exploits only the search space visible to its monogamous parents using mainly the crossover operator. As offspring are "protected" by their parents in this mode, relatively weaker offspring in view of the population have as good chance a of survival as other stronger offspring from the other families. Consequently, we do not observe any drastic changes to the population diversity at this stage. However, once the bonds are broken, selection pressure increases and forces every individual to participate in the mate selection process in order to form new monogamous pairs. The newly formed population consists of better fit individuals but may at the same time be less genotypically similar.

10.4.4. *Computational effort*

Moving on, this section discusses the computational effort chalked by each algorithm. The actual computational time is taken into consideration, where each algorithm is allowed to execute to completion, i.e. when maximum iteration is reached in all trial runs. Here, the time complexity of each algorithm is take into consideration, rather than its convergence speed. Figure 10.8 illustrates the empirical results.

Generally, all the three algorithms spend most of their processing time on fitness evaluations. In comparison to the SGA, the evaluation phase is greater in monogamous GAs. Such can be explained by the mandatory crossover operations ($P_c = 1$) in both algorithms, which in turn requires an offspring fitness evaluation for every new breed. This also explains why the percentage of crossover operation required in monogamous GAs is higher than that of SGA. In this example, the SGA would spend theoretically 85% of its evolutionary process (note that this does not map directly to *computational time*) in crossover since $P_c = 0.85$. Inherently, it also has profound impact on ensuing operations such as the mutation, since mutation proceeds crossover in the breeding stage.

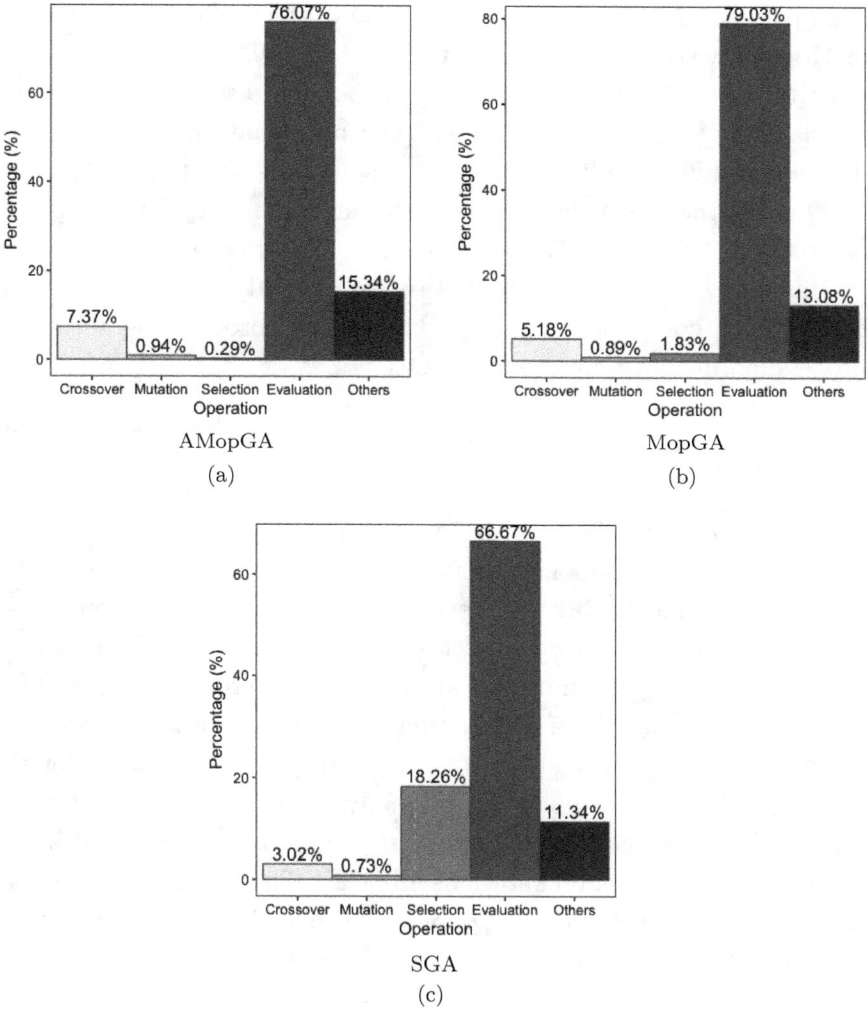

Fig. 10.8. Pie charts of operation time.

The selection operation consumes the second highest processing time in SGA. On the contrary, the selection operation has been greatly reduced in both monogamous GAs (less 2% of the overall processing time). In SGA, parents must be selected at each generation. In contrast, selection only occurs at selective generations (during rebond) in the other two algorithms. To give an immediate example, only 22% (= (0.02+0.2)×100%) of the evolutionary process of MopGA with parameters $g_p = 0.02$ and $P_i = 0.2$ is spent in selecting the parents. Therefore, there is economical use of the selection mechanism in monogamous GAs. Take note that this is the most distinctive difference between monogamous GAs and the SGA.

Overall, there is a slight increase in mutation usage in AMopGA compared to MopGA and SGA. The latter are fixed at $\frac{10}{L}$, for instance, while the AMopGA is adapted dynamically. Lastly, the "other" components in the evolutionary process included IO processing, memory management, etc.

10.5. Conclusion

Breaking the norm of classical GAs, this chapter describes a variant of GA that promotes monogamous breeding. Known as monogamous parents, the same mating pair performs crossover and mutation operations at every generation until the pair bond tenure is reached. It has been shown that multiple applications of crossover operation lead to higher chances of discovering better fit offspring. Generally, the notion is to extensively exploit the current search space prior to exploring the other regions. Additionally, in order to promote population diversity, monogamous parents are allowed to mate with an extrapair, a phenomenon known as infidelity. This operation not only insures that a healthy level of population diversity is sustained, but also spreads genetic materials across the population, leading to faster convergence.

Two MopGA models have been introduced — a standard MopGA, followed by an adaptive model, AMopGA. All variables, except the population size, are dynamically or adaptively modified during the evolutionary process in AMopGA. The models have been empirically proven to outperform the SGA in terms of computational effort as well as solution quality.

We are only beginning to grasp the mystery of monogamy, whether naturally or artificially. More extensive theoretical and/or empirical analyses are necessary in the spirit of Science. Nonetheless, it is hopeful that monogamous GAs will constitute towards the direction of highly effective evolutionary algorithms.

Bibliography

[1] Botero, C. A. and Rubenstein, D. R. Fluctuating environments, sexual selection and the evolution of flexible mate choice in birds. *PLoS One* **7**(2), e32311, 2012.

[2] Culberson, J. C. Genetic invariance: A new paradigm for genetic algorithm design, Tech. Rep., pp. 92–106, 1992. Department of Computing Science, University of Alberta.

[3] Culberson, J. C. Mutation-crossover isomorphisms and the construction of discriminating functions. *Evol. Comput.* **2**(3), 279–311, 1994.

[4] De, S., Pal, S. K. and Ghosh, A. Genotypic and phenotypic assortative mating in genetic algorithm. *Inf. Sci.* **105**(1), 209–226, 1998.

[5] De Jong, K. A. and Spears, W. M. A formal analysis of the role of multipoint crossover in genetic algorithms. *Ann. Math. Artif. Intell.* **5**(1), 1–26, 1992.

[6] Forrest, S. and Mitchell, M. Relative building-block fitness and the building-block hypothesis. *Found. Genet. Algorithms* **2**, 109–126, 1993.

[7] Goldberg, D. E. *Genetic Algorithms in Search, Optimization and Machine Learning*, 1st edn. Addison-Wesley Longman Publishing Co., Inc., Boston, MA, 1989.

[8] Griggio, M. and Hoi, H. An experiment on the function of the long-term pair bond period in the socially monogamous bearded reedling. *Anim. Behav.* **82**(6), 1329–1335, 2011.

[9] Hatchwell, B., Russell, A., Ross, D. and Fowlie, M. Divorce in cooperatively breeding long-tailed tits: a consequence of inbreeding avoidance? *Proc. R. Soc. Lond. B, Biol. Sci.* **267**(1445), 813–819, 2000.

[10] Holland, J. Adaptation in natural and artificial systems: an introductory analysis with applications to biology. *Control Artif. Intell.* 1975.

[11] Jie, L., Liu, W., Sun, Z. and Teng, S. Hybrid fuzzy clustering methods based on improved self-adaptive cellular genetic algorithm and optimal-selection-based fuzzy c-means. *Neurocomputing* **249**, 140–156, 2017.

[12] Kurdi, M. An effective new island model genetic algorithm for job shop scheduling problem. *Comput. Oper. Res.* **67**, 132–142, 2016.

[13] Lim, T. Y., Al-Betar, M. A. and Khader, A. T. Adaptive pair bonds in genetic algorithm: An application to real-parameter optimization. *Appl. Math. Comput.* **252**, 503–519, 2015.

[14] Lim, T. Y., Al-Betar, M. A. and Khader, A. T. Monogamous pair bonding in genetic algorithm. In *2015 IEEE Congress on Evolutionary Computation (CEC)*, pp. 15–22, 2015.

[15] Lim, T. Y., Al-Betar, M. A. and Khader, A. T. Taming the 0/1 knapsack problem with monogamous pairs genetic algorithm. *Expert Syst. Appl.* **54**, 241–250, 2016.

[16] Mitchell, M., Forrest, S. and Holland, J. The royal road for genetic algorithms: Fitness landscapes and GA performance. In *Proceedings of the First European Conference on Artificial Life.* Cambridge: The MIT Press, pp. 245–254, 1992.

[17] Ramsay, S. M., Otter, K. A., Mennill, D. J., Ratcliffe, L. M. and Boag, P. T. Divorce and extrapair mating in female black-capped chickadees (parus atricapillus): separate strategies with a common target. *Behav. Ecol. Sociobiol.* **49**(1), 18–23, 2000.

[18] Resendez, S. L. and Aragona, B. J. Aversive motivation and the maintenance of monogamous pair bonding. *Rev. Neurosciences* **24**(1), 51–60, 2013.

[19] Rudolph, G. Convergence analysis of canonical genetic algorithms. *IEEE Trans. Neural Netw.* **5**(1), 96–101, 1994.

[20] Schuiling, G. The benefit and the doubt: why monogamy? *J. Psychosom. Obstet. Gynecol.* **24**(1), 55–61, 2003.

[21] Spears, W. M. Crossover or mutation? In *Foundations of Genetic Algorithms 2 (FOGA-92)*, pp. 221–237, 1992.

[22] Tan, C. J., Lim, C. P. and Cheah, Y.-N. A modified micro genetic algorithm for undertaking multi-objective optimization problems. *J. Intell. Fuzzy Syst.* **24**(3), 483–495, 2013.

[23] Tan, C. J., Neoh, S. C., Lim, C. P., Hanoun, S., Wong, W. P., Loo, C. K., Zhang, L. and Nahavandi, S. Application of an evolutionary algorithm-based ensemble model to job-shop scheduling. *J. Intell. Manuf.* 1–12, 2017. doi 10.1067/s10845-016-1291-1. https://doi.org/10.1007/s10845-016-1291-1.

[24] Watson, R. A. and Jansen, T. A building-block royal road where crossover is provably essential. In *Proceedings of the 9th Annual Conference on Genetic and Evolutionary Computation (ACM)*, pp. 1452–1459, 2007.

Index

www.ingramcontent.com/pod-product-compliance
Lightning Source LLC
LaVergne TN
LVHW010333220926
848484LV00004BB/890

* 9 7 8 9 8 1 3 1 4 3 1 7 3 *